CEREBRUM 2017

Cerebrum 2017

Emerging Ideas in Brain Science

Bill Glovin, Editor

DANA PRESS

New York

Published by Dana Press, a Division of the Charles A. Dana Foundation, Incorporated

Address correspondence to:
Dana Press
505 Fifth Avenue, Sixth Floor
New York, NY 10017

THE
DANA
FOUNDATION

New York, NY 10017
DANA is a federally registered trademark.
Printed in the United States of America
ISBN-13: 978-1-932594-63-8
ISSN: 1524-6205
Book design by Bruce Hanson at EGADS (egadsontheweb.com)
Cover illustration by William Hogan

Interior illustrations:
Page 4—William Hogan
Page 15—William Hogan
Page 26—Shutterstock
Page 36—Seimi Rurup
Page 46—Shutterstock
Page 58—Shutterstock
Page 68—Seimi Rurup
Page 78—William Hogan
Page 92—Shutterstock
Page 103—Seimi Rurup
Page 130—William Hogan
Page 142—Seimi Rurup

CONTENTS

The book reflects Fins' role as co-director of the Consortium for the Advanced Study of Brain Injury at Weill Cornell Medicine and the Rockefeller University and his struggle to answer the kinds of questions that stand to shape how society treats people with brain injuries. What is the capacity of brains to recover? What are the mechanisms of that recovery? How do we know that our assessments are accurately describing what's going on in a patient's mind? And what does society morally owe these patients and families?

Review by Arthur L Caplan, Ph.D.

A primary function of my role as editor is asking top neuroscientists to write about the latest developments in their specialty areas for lay readers. If they agree to the assignment, I encourage them to use—whenever possible—conversational language, anecdotes, storytelling, and their own voice in communicating what are often complex and hard-to-explain topics. Another option might be to suggest they read Alan Alda's new book before they begin.

Review by Eric Chudler, Ph.D.

Buddhism shares with science the task of examining the mind empirically. But Buddhism has pursued, for two millennia, direct investigation of the mind through penetrating introspection. Neuroscience, on the other hand, relies on third-person knowledge in the form of scientific observation. In the book that is the subject of this review, two friends, one a Buddhist monk trained as a molecular biologist, and the other, a distinguished neuroscientist, offer their perspectives on the mind, the self, consciousness, the unconscious, free will, epistemology, meditation, and neuroplasticity.

Review by Paul J. Zak, Ph.D.

Foreword

Neuroscience Beyond the Nervous System

by Ed Bullmore, Ph.D.

Ed Bullmore, Ph.D., is head of the Department of Psychiatry and director of the Wolfson Brain Imaging Centre in the University of Cambridge and director of research and development in Cambridgeshire & Peterborough Foundation NHS Trust. Since 2005, he has worked half-time for GlaxoSmithKline, currently focusing on immuno-psychiatry. Bullmore trained in medicine at the University of Oxford and St Bartholomew's Hospital, London; then in psychiatry and MRI at the Bethlem Royal & Maudsley Hospital, London. His work on brain network science and brain imaging has been highly cited.

IN THE 20TH CENTURY, neuroscience took root and was firmly grounded in understanding the nerve cell. This was the right place to start. We needed to see nerve cells and the synaptic connections between them precisely, to begin piecing together how neuronal networks might do information processing and how the brain might generate moods, thoughts, and other facets of our subjective mental experience.

In the 21st century, as the *Cerebrum Anthology 2017* is sign-posting, one key direction of travel for neuroscience will be beyond the nerve cell—indeed beyond the nervous system.

Among the articles is "Microglia: The Brain's First Responders," which highlights a vivid example of this shift in focus. This article provides a clear, succinct, and balanced introduction to microglial cells and their potential therapeutic applications. I can recommend it as an excellent primer for readers who are thoroughly familiar with the idea that the brain is full of nerve cells but not so well acquainted with the notion that it contains a lot of immune cells as well.

When the 19th century pioneers of neuroscience first peered down their brass microscopes, they saw a complex web of neural processes along with diverse other cells. The constitution of the neural web attracted the lion's share of interest and controversy, as signified by the joint award of the 1906 Nobel Prize to Santiago Ramón y Cajal and Camillo Golgi for their equally brilliant but contradictory theories about the nerve cell. Other brain cells were relatively neglected and assumed to have a lowly function, mostly as physical support structures to stabilize the nervous system. Microglia means literally small glue cell, and that, traditionally, is all they were thought to do—glue the nerve cells together.

As **Staci Bilbo** and **Beth Stevens** elegantly explain, we now know that microglia are anatomically located in the brain but functionally part of a large family of immune cells that include macrophages and monocytes and are distributed widely throughout the body. We are beginning to see from many exciting angles how this close juxtaposition of nerve and immune cells is highly relevant to the function, development, and disorders of the brain.

What do microglia do? Like macrophages elsewhere in the body, they eat things—specifically, things they have identified as damaging. For example, if your brain is infected with bacteria, as the first line of defense microglial cells will rapidly envelop, kill, and digest them to help you survive. But microglia can eat many things besides infectious bacteria, such as synapses and amyloid protein plaques. This makes them constructive partners in the developmental formation of the nervous system, but it means they can also play a destructive role, mediating toxic inflammatory processes that accelerate the neurodegenerative changes of Alzheimer's disease. They are, in short, much more important than their name suggests.

The new science of neuro-immunology, exemplified by this fresh respect for the formerly humble microglial cell, has fascinatingly diverse and disruptive implications for how we think about the brain and its disorders. We can hear other echoes of communication between the nervous system and the immune system in this year's anthology.

The human brain takes a long time after birth to mature. As **Nim Tottenham** lucidly explains in "The Brain's Emotional Development," this is especially true of brain circuits for emotion and cognitive control of emotion. In animals, as in humans, emotions depend on a large-scale network of synaptically inter-connected brain regions. A key interaction within this network is between the amygdala, a subcortical region for emotional learning, and the medial prefrontal cortex, a cortical region for emotional regulation. A series of elegant studies in animals and humans have shown connectivity between these areas to have a prolonged and distinctly phased course of post-natal development.

Connectivity from the amygdala to the medial prefrontal cortex forms relatively early, between weaning and puberty, whereas the controlling negative feedback connections from the medial prefrontal cortex to the amygdala form later, between puberty and adulthood. This sequence of developmental phases, or critical windows, is consistent with the dependence of pre-pubertal children on parental guidance to help them control emotions that they are unable to control themselves. It is also compatible with the high risk of depressive and other psychiatric symptoms in adolescents, who must navigate a path to emotional maturity less dependent on parental support but not yet supported by late-developing adult-emotion control systems.

What does this have to do with microglial cells and the immune system? The connection is as yet tenuous but intriguing. Developmental progression from a "childish" state of emotional brain network organization to a more "grown-up" mode must depend on a massive program of synaptic re-modeling and pruning. And that, as cited above, is one of the things that microglia do. Besides identifying and eliminating redundant synapses, they trophically encourage the formation of new ones. Microglia could thus be change agents in emotional brain development.

If so, then one would predict that environmental shocks or genetic variations associated with altered microglial activation might put normal emotional brain development at risk and increase vulnerability to mood disorders. There is some circumstantial evidence in support of this prediction. Stress, a major environmental risk factor for depression, is increasingly recognized to be a potent cause of microglial activation and peripheral inflammation as well. As highlighted by Tottenham's article, one of the ways that parents help their children reach maturity and independence is by controlling the biological stress of early emotional experiences, e.g., by modulating the child's production of anti-inflammatory steroids in response to emotional threats. So one can speculate that good parenting benefits emotional development by protecting the normal microglial program of synaptic remodeling against perturbation by stress-induced inflammatory shocks.

As a psychiatrist, I find this pathogenic hypothesis deeply interesting because of the way it promises to bring together what we have long known about social and psychological risks for depression with newer knowledge about emotional brain development and the microglial mechanisms underlying synaptic plasticity. It goes beyond the nervous system. In that sense it is akin to recent work on the genetics of schizophrenia that has identified a gene for an inflammatory protein, C4, as a controlling factor in synaptic pruning and a risk factor for psychotic disorder.

"Gut Feelings on Parkinson's and Depression," by **Ted Dinan** and **John Cryan**, provides an equally disruptive perspective on the pathogenesis of Parkinson's disease, traditionally considered to be entirely a disease of the nervous system. They start by reminding us that although we usually think of ourselves as a single organism, each of us actually represents an ecosystem. Our gastro-intestinal tracts are home to a dense and diverse community of other species, the bacteria that constitute the microbiota or microbiome. We have known about intestinal bacteria for decades but, like the little glue cells in the brain, we have traditionally underestimated their importance for the health of the nervous system.

Dinan and Cryan explain that intestinal bacteria help synthesize neurotransmitters, like GABA and dopamine, and stimulate the production of gut hormones, like leptin and ghrelin, that act on the brain to control appetite and food consumption. There are, in fact, many channels of communication between the microbiome, the gut, and the brain. The vagus nerve innervates the gut wall extensively, and preliminary results of research suggest this extraordinary hypothesis: brain deposits of the protein alpha synuclein, characteristic of Parkinson's disease, are produced by the microbiome and transported to the brain via the vagus nerve.

If this pathogenic hypothesis is correct, you might expect that the microbiome should be constituted differently in patients with Parkinson's disease compared to healthy controls. There is, in fact. preliminary data that the microbiome is sufficiently abnormal in Parkinson's disease that patients can be distinguished from controls by analysis of fecal samples. The hypothesis would also predict that surgical interruption of the vagus nerve, by a procedure called vagotomy, should reduce the risk of Parkinson's disease. And there is retrospective evidence that patients who were treated with vagotomy for peptic ulcer (before the advent of histamine antagonists made surgical treatment unnecessary) have reduced incidence of Parkinson's disease.

Vagal transport of alpha-synuclein from the microbiome to the brain is not the only way the bowel's inhabitants might influence neuropsychiatric disease. Macrophages of the immune system densely populate the lining of the gastrointestinal tract and closely monitor their local environment for invasion of the gut wall by hostile bacteria from the microbiome. Activated by such bacteria, these immune cells produce inflammatory cytokines that circulate in the blood and communicate across the blood brain barrier to activate microglia. In other words, the immune system can feed information to the brain concerning the status of the microbiome. This channel of communication could also be relevant for Parkinson's disease. Genetically predisposed mice are at increased risk of developing neurological symptoms if they are reared under dietary conditions that stimulate microglial activation; and, conversely, are somewhat protected if they are treated with minocycline, an anti-inflammatory antibiotic drug.

Immune-mediated communication between the gut and the brain might also contribute to the pathogenesis of depression. There is now robust evidence that a proportion of patients with major depressive disorder (MDD) have increased levels of cytokines in circulation, and so-called "leaky gut syndrome"—the infiltration of the gut wall by hostile bacteria—is a plausible source of such peripheral inflammation.

One very recent result that seems consistent with these emerging ideas comes from large-scale genome wide association studies (GWAS) of MDD that have finally achieved sufficient statistical power to identify risk genes for depression robustly. A meta-analysis by the Psychiatric Genetics Consortium suggests that many of these genes have functional roles in the immune system as well as the nervous system. Indeed, the top hit was a gene called olfactomedin 4 that was previously known to be important in controlling the inflammatory reaction of the stomach wall to infection with *H. pylori,* a hostile agent in the community of the microbiome.

So we can begin to think beyond the nervous system as we try to understand the causes of mood and neurodegenerative disorders. But what can this new perspective offer up in terms of new treatments?

We all know that there is as yet no cure for Alzheimer's disease; but that is not to say there is nothing to be done about it. As someone who is now closer than he'd like to the age range of greatest risk for Alzheimer's disease onset, I naturally welcomed the optimistic review article by **Dharma Singh Kalsa** and **George Perry** on "The Four Pillars of Alzheimer's Prevention." They summarized recent research in support of four interventions that could make a difference in age-related decline of cognitive performance: diet, exercise (physical and mental), meditation, and psychological well-being. For each of these factors there is empirical evidence of benefit to the aging brain and mind. The evidence base in support of kirtan kriya meditation is particularly interesting. This ancient technique of mindfulness requires just 12 minutes practice daily but has improved cognitive outcomes in several recent studies. As the authors explain, it is becoming clearer how kirtan kriya and other practices that improve psychological well-being might have beneficial effects. Besides controlling stress, meditation is associated with changes in inflammatory gene expression and in MRI measures of brain structure and function in components of the emotional brain network and elsewhere.

You can begin to see why there is increasing interest in the development of new drugs, for the treatment of both Alzheimer's disease and depression, that might work primarily on the immune system rather than the nervous system. While effective anti-depressant drugs such as serotonin reuptake inhibitors (SSRIs) are already available, many patients respond incompletely to them, and there have been no major treatment innovations in the last 20 years. For the therapeutic progress that I believe is long overdue, we will need to move beyond the "old school" approach of trying to find a single drug, a panacea that will work equally well for all depressed patients. Psychiatry generally must take a more personalized approach, using blood biomarkers and brain imaging to match individual patients to the treatment options most likely to work for them.

Boadie Dunlop and **Helen Mayberg's** review of "Neuroimaging Advances for Depression" is a nuanced report on the state of the art. As they explain, several brain scanning methods used in clinical studies of depression have yielded reports of abnormal structure or function, especially of the prefrontal, amygdala, and cingulate regions that comprise the emotional brain network. However, when the MRI literature is examined more rigorously, the evidence for a consistent difference between MDD cases and healthy controls is less than compelling. This might be because the clinical heterogeneity of MDD is associated with similarly heterogeneous brain changes whose inter-patient variability is too high to detect a consistent case-control difference. But the picture is no clearer when the clinical phenotype is refined by sub-typing patients with, for example, melancholic or psychotic depression.

Either standard psychiatric nosology is not carving nature at the joints, or there are no brain changes in depression that are detectable by MRI.

Dunlop and Mayberg argue in favor of the former, pointing out that when data are analyzed not through the prism of clinical diagnosis but by a bottom-up, data-driven approach, the evidence connecting imaging markers to depressive symptoms becomes more compelling. They also make a strong case that MRI could play a useful role in predicting response to drug and other treatment, rather than serving simply as a mirror of clinical states. It will be interesting to see whether MRI can be further developed as a proxy of brain inflammatory states predictive of a poor response to conventional anti-depressant drugs or a therapeutic response to novel anti-inflammatory interventions.

But is there a cost to going beyond the nervous system? Do we jeopardize the reductionistic focus on the nerve cell that has successfully driven so much of 20th century neuroscience by broadening our gaze to encompass immune cells? I was reassured by an observation from Eric Kandel's masterly book, *Reductionism in Art and Brain Science: Bridging the Two Cultures*, reviewed by **Ed Bilsky**: "often critical insights are gained by combining approaches…a major step forward in the study of the brain was the scientific synthesis that occurred in the 1970s, when psychology, the science of the mind, merged with neuroscience, the science of the brain." It made me wonder if we are now moving towards a new synthesis of cognitive neuroscience, the science of the mind/brain, with immunology, the science of the self and the non-self.

Going beyond the nervous system, thinking about the links between brain and body and those between the brain and the outside world, many of them mediated by the immune system, was the thread I followed through the *Cerebrum* anthology. I am sure you will enjoy following your own thread.

ARTICLES

1

Examining the Causes of Autism

By David G. Amaral, Ph.D.

David G. Amaral, Ph.D., is a Distinguished Professor in the Department of Psychiatry and Behavioral Sciences at UC Davis. He is also the Beneto Foundation Chair and Research Director of the MIND Institute, which is dedicated to studying autism and other neurodevelopmental disorders. As research director, he coordinates a multidisciplinary analysis of children with autism called the Autism Phenome Project to define clinically significant subtypes of autism. More recently, Amaral has become Director of Autism BrainNet, a collaborative effort to solicit postmortem brain tissue to facilitate autism research. In April of 2015, Amaral became editor-in-chief of Autism Research, the journal of the International Society for Autism Research. In 2016, he was appointed to the Interagency Autism Coordinating Committee by the Secretary of Health and Human Services. Amaral received a joint Ph.D. in neuroscience and psychology from the University of Rochester and conducted postdoctoral research at the Department of Anatomy and Neurobiology at Washington University. He also conducted research at the Salk Institute for Biological Studies and served as an adjunct professor in the Department of Psychiatry at UC San Diego.

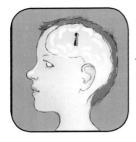

Editor's Note: Autism is a broad, complex, and increasingly important brain disorder. New data from the Center for Disease Control and Prevention indicate that one in sixty-eight children is born with some degree of autism. Autism is also more common in males by a four to one ratio. Making it especially difficult to discuss in finite, conclusive terms is the fact that there is no biological test for autism; diagnosis is based on behavior, and the only verified treatment is intensive behavior therapy. Our author, one of the nation's foremost researchers on autism, examines the prenatal factors that contribute to the disorder.

AS AN AUTISM RESEARCHER, I often try to put myself in the shoes of parents who have just been told that their child has autism. More and more families in the United States and around the world are facing this difficult news. The families that I've seen go through this often respond emotionally at first. Some go through denial; others are sad or furious. But emotions soon give way to questions. What caused my child's autism? Was I to blame? Which treatments will help? And what does the future hold?

Autism research has made tremendous progress over the last 20 years, but yet we still can't provide definitive answers to most of these questions. I find the autism community to be proactive, combative, and opinionated. The complexity and ambiguity of autism has spawned myriad speculations about causes—many of which have little supportive evidence. It seems clear at this point, however, that when all is said and done, we will find that autism has multiple causes that occur in diverse combinations.

To begin with, many people struggle to understand the nature of a condition so wide ranging in its severity. Autism Spectrum Disorder (ASD) or autism is a behaviorally defined neurodevelopmental disorder characterized by 1) persistent deficits in social communication and interaction across multiple contexts, and 2) restricted, repetitive patterns of behavior, interests, or activities. Few would dispute that the causes of ASD include both genetic and environmental factors. Indeed, more than 100 genes are known to confer risk[1,2] and 1,000 or more may ultimately be identified.[3] A wide range of potential environmental challenges have also been associated with autism, although studies in this area lag behind genomics research. A short overview of data supports genetic and environmental contributions to ASD etiology. A focus on prenatal events will hopefully clarify that the cause of autism, in the vast majority of cases, occurs prenatally, even if behavioral signs first appear several years after birth.

Twin Studies

Strong evidence against the unfounded view that autism results from neglectful parenting came in 1977 from Folstein and Rutter and the first systematic, detailed study of twin pairs containing at least one child with autism.[4] In this study, 11 of the twin pairs were monozygotic (nearly identical genetics) and 10 were dizygotic (shared approximately half of their genome with each other). The major finding was that four of the monozygotic twin pairs were concordant (both had autism), whereas none of the dyzygotic twins were. Beyond autism, nine of the eleven monozygotic pairs were concordant for some form of cognitive impairment, compared to one of ten of the dyzygotic pairs.

The researchers concluded that autism and other neurodevelopmental disorders have a strong genetic component. Environmental factors must also contribute to autism etiology. For the 17 twin pairs that were discordant for autism—one child had a diagnosis and the other did not—the authors speculated that direct damage to the brain might have affected the diagnosed twin. They identified five features known to be associated with brain damage, such as severe hemolytic disease, a delay in breathing of at least five minutes after birth, and neonatal convulsions. In six of the pairs, one twin—always the autistic one—experienced one or more of these insults. Looking further, they found that one of an expanded list of "biological hazards" (e.g., discrepancies in birth weight, a pathologically narrow umbilical cord) occurred in the autistic twin in 6 of the 11 remaining discordant pairs and never in the non-autistic twin. The authors concluded that "some form of biological impairment, usually in the perinatal period, strongly predisposed to the development of autism."

Since the Folstein and Rutter paper cited above, there have been a total of 13 twin studies focused on autism. All find genetic and environmental contributions to autism, although conclusions about the proportions of the two factors and interpretations have varied substantially. One research team,[5] for example, concluded that a large proportion of the variance in liability (55 percent for strictly defined autism and 58 percent under a broader definition) can be explained by shared environmental factors, whereas genetic heritability accounts for 37 percent. This somewhat surprising finding—that environmental factors contribute more substantially than genetics—has been challenged by a more recent, large-scale twin study,[6] which found that the largest contribution to autism liability comes from additive genetic effects. And, a recent meta-analysis[7] concludes that the causes of autism are due to strong genetic effects, and that shared environmental influences are seen only if autism is very narrowly defined. A brief synopsis of the history of autism twin studies[8] finds that concordance for monozygotic twins is roughly 45 percent, versus 16 percent for dizygotic twins.

The reason for this short review of autism twin studies is to emphasize that even the best evidence for both genetic and environmental etiologies of autism leads to inconsistent conclusions about their proportional contributions. Moreover, twin studies do not typically consider that the cause of autism may involve   genetic and environmental factors working together (the so-called gene by environment effect); i.e., certain environmental exposures only cause autism in individuals with a particular genetic composition. The second point is that if autism had a completely genetic etiology, we would expect a much higher concordance rate in monozygotic twins; the actual rate may reflect, in part, that even monozygotic twins do not share an identical environment prenatally.[9,10] Therefore, one must seriously search for environmental factors that either alone, or in combination with genetic predisposition, can increase autism risk. What are these factors?

Maternal Infection

If twin studies provide the best evidence for a genetic basis of autism, then naturally occurring pathogen exposures offer the strongest evidence of environmental etiology. The best example is maternal rubella (German measles) infection during pregnancy. Before development and widespread dissemination of effective vaccines, major pandemics occurred every 10 to 30 years.[11] The last of these was from 1963 to 1965 and infected an estimated 10 percent of pregnant women, resulting in more than 13,000 fetal or early infant deaths; 20,000 infants born with major birth defects and 10,000 to 30,000 infants born with moderate to severe neurodevelopmental disorders. Stella Chess, a child psychiatrist at New York University, studied 243 children exposed to rubella during pregnancy[12,13] and found that the largest category of neurodevelopmental disorder was intellectual disability, which affected 37 percent of the sample. Nine of these children were also diagnosed with autism; another, without intellectual disability, had a possible diagnosis; and eight a partial syndrome of autism. These numbers would translate to an autism prevalence of 741 per 10,000 rubella-exposed children, just over seven percent. This is striking in comparison to published prevalence rates, at the time of the study, of two to three per 10,000 in the general population. Fortunately, rubella epidemics have ended due to widespread dissemination of the measles, mumps and rubella vaccines and the association of autism with other viral or bacterial infections is weaker than with rubella.[14]

Collier et al[15] have pointed out that nearly 64 percent of women surveyed in the US have experienced an infection during their pregnancies. This obviously does not lead to autism or any other neurodevelopmental disorder in most cases.

Examining prenatal environmental factors is best conducted in very large cohorts of subjects that have excellent health care records. This can be done in Scandinavian countries with their nationalized health care systems, and in large health care providers in the US.

One such study, conducted in Denmark, found no association between maternal bacterial or viral infection during pregnancy and diagnosis of ASD in the offspring,[16] although viral infection during the first trimester, or admission to the hospital due to infection during the second trimester were associated with the diagnosis. In a more recent study[17] Atladottir and colleagues found little evidence, overall, that common infectious diseases or fevers (lasting more than seven days) during pregnancy increased the risk of autism—noting, however, that influenza increased the risk of having an autistic child twofold. Use of antibiotics also increased risk. The link between influenza exposure during fetal life and increased risk for autism is in line with a series of animal studies[18, 19] suggesting that the influenza virus activates the maternal immune system, which may be harmful to fetal brain development. But the Danish researchers seem to downplay even their statistically significant findings, suggesting that their results do not indicate that either mild infection or the use of antibiotics represent strong risk factors for autism.

A parallel set of studies has been carried out by Zerbo and colleagues in California. The first,[20] based on 1,122 children, found no association between maternal influenza and ASD but (in contrast to Atladottir et al), the occurrence of maternal fever did increase risk. A second study[21] of 2,482 children (407 with ASD) found that mothers of children with ASD were diagnosed with viral infections during pregnancy no more often than mothers of non-autistic children. Maternal bacterial infections during the second trimester and the third trimester, however, were associated with a twofold increase in ASD risk, and two or more infections diagnosed in the third trimester with even higher risk, again suggesting a link with more severe infection during pregnancy. The most recent study,[22] based on a large cohort of children (196,929) born between 2000 and 2010, found that neither maternal influenza infection during pregnancy nor influenza vaccination were associated with increased risk for ASD.

In conclusion: Some infections during pregnancy, such as German measles, clearly increase the risk of ASD. However, there seems relatively little evidence that today's widely experienced infectious illnesses, such as influenza, during pregnancy substantially increase the risk of ASD. Perhaps the signal is weak because of gene by environment effects [as seems to be the case for different strains of mice[23, 24]]. If so, evidence will need to come from studies that combine large scale epidemiology with sophisticated genomic analyses.

Maternal Antibodies

Autoimmune diseases (in which immune cells erroneously identify cells in the body as foreign and attack them) mediated by circulating antibodies currently affect as much as nine percent of the world's population,[25] and the notion that autoimmunity may be associated with neurological and psychiatric disorders goes back to the 1930s. Reviewing this contentious area of research, Goldsmith and Rogers[26] conclude that the literature, though conflicting, "contains a large amount of circumstantial, but not conclusive, evidence for immune dysfunction in patients with schizophrenia." Interestingly, an auto-immune disorder with antibodies directed at the NMDA receptor causes an encephalopathy, which in its early stages can be indistinguishable from schizophrenia.[27]

Precedents for antibody-related CNS disorders include Rasmussen encephalitis, stiff-person syndrome, neuromyelitis optica, post streptococcal movement disorders (Sydenham's chorea and PANDAS), and systemic lupus erythematosus.[28] Judy Van de Water, of UC Davis, the main proponent of the idea that circulating antibodies may cause some forms of autism, first reported in 2008 that 12 percent of mothers of children with ASD have unusual antibodies directed at fetal brain proteins.[29] Based on more specific assays for these antibodies, she has since proposed that Maternal Antibody-Related (MAR) causes may be associated with as many as 22 percent of autism cases, suggesting that this may be a preventable form of ASD.[30] This area of research is exciting because it suggests potential therapeutic targets. Although many questions remain (e.g., how antibodies would enter the fetal brain, what neurodevelopmental processes they may alter), it is entirely possible that circulating antibodies represent prenatal environmental risk factors for ASD.

Drugs

Efforts to understand the increased prevalence of autism spectrum disorder have led some to wonder whether the use of various drugs during pregnancy might be partly responsible. Historically, a strong case could be made for an association between autism and thalidomide, a potent sedative that was used (for several years around 1960) during pregnancy for the relief of nausea. A study of 100 adult Swedish patients whose mothers had taken thalidomide while pregnant[31] found that at least four had clear autistic characteristics. This was the first evidence that a drug ingested during pregnancy could substantially increase autism risk. More recently, concerns have been raised about valproic acid and serotonin reuptake inhibitors.

Valproic acid, an approved drug since the early 1960s, is primarily prescribed for epilepsy and seizure control, but also used for ailments ranging from migraine headaches to bipolar disorder. Both animal and human epidemiological studies have raised concerns that valproic acid is a teratogen. The largest epidemiological study to date[32] tracked 415 children, 201 of whom were born to mothers who took antiepileptic medication during their pregnancies. Nearly 7.5 percent of the children of the treated women had a neurodevelopmental disorder, primarily some form of autism, versus 1.9 percent in the non-epileptic women.

A recent concern has been the use of serotonin reuptake inhibitors (SSRIs) for the treatment of depression during pregnancy. Serotonin is an important brain neurotransmitter that plays a significant role in functions ranging from sleep to mood to appetite, and whose dysregulation during early fetal life can have serious negative consequences for brain development.[33] As the name implies, SSRIs, which have been in use since the late 1980s, delay the reuptake of serotonin from the synaptic cleft into the presynaptic terminal and thus enhances its effect on the postsynaptic receptors. A recent review and meta-analysis of six case-control studies and four cohort studies concluded that SSRI use during pregnancy[34] was significantly associated with increased risk of ASD in offspring.

The effect was most prominent with use of the drugs during the first and second trimesters of pregnancy. Interestingly, the researchers found that preconceptual exposure to SSRIs was also associated with increased ASD risk—as was the use of non-SSRI antidepressants. They note that a large cohort study found that, while ASD rates in the SSRI-exposed group were significantly higher than in the unexposed group, the rates in the SSRI-exposed group did not significantly differ from those among mothers with unmedicated psychiatric disorder and those who had discontinued SSRIs. It currently appears impossible to disentangle the deleterious effect of SS-RIs from the fact of a maternal condition that necessitates the drug. Many authors also comment on the potentially worse effect on pregnancies of untreated maternal depression.

In sum, a brief review of the literature indicates that ingesting some drugs during pregnancy increases the risk of ASD, suggesting the need for more careful evaluation of drug safety during fetal development prior to widespread medical use.

Environmental Toxicants

Beyond viral and bacterial pathogens and medically prescribed drugs, researchers have begun investigating environmental toxicants. These range from automobile-produced air pollution to cigarette smoke to heavy metals and pesticides.[35,36] Small increases in autism risk have been reported if, for example, a family lives closer to a freeway or to an agricultural area during pregnancy. The field of autism environmental epidemiology is still in its infancy and techniques to comprehensively establish a prenatal "exposome" (i.e., all environmental factors affecting a fetus during pregnancy) are still under development. That said, given the unlikelihood that all autism will be explained by genetic factors, the determination of environmental causes, some of which might be avoided or minimized, may have far greater translational impact than the much better funded genetic studies. Strategies for exploring gene-by-environment interactions need to be enhanced with haste.

Postnatal Factors

Since autism is a neurological disorder that undoubtedly reflects altered brain function, it is possible that the insult to the brain occurs after birth. There is currently very little evidence for this. One historical concern was that vaccines, such as the measles, mumps, and rubella (MMR) vaccine, administered initially when the child is about one-year old, might transform a healthy child into one with autism. This fear was fueled by regressive onset in some cases—a child seems fine for the first year or so, then loses social and language function and regresses into a classical autistic syndrome. But we have found that even in children who demonstrate this regressive form of autism, brain changes begin by four to six months, long before behavior changes.[37] Moreover, many large-scale epidemiologic studies have unequivocally demonstrated no link between MMR administration and the risk of ASD (summarized in [38]), the same conclusion that the US National Academy of Sciences reached in a thorough review carried out in 2011.[39]

The only other postnatal experience that has been linked to the onset of ASD is profound social isolation in institution-reared children, such as those in the Romanian orphanage system.[40] Rutter and colleagues[41] found that nearly 10 percent of children raised in Romanian orphanages and adopted by British families showed some features of autism. These children were very poorly treated in the orphanage (most were underweight and had intellectual disability and various medical problems). While fully qualifying for an autism diagnosis at age 4, they showed substantial improvement and less severe autism symptoms by age 6. Is this truly autism? The authors conclude: "The characteristics of these children with autistic features, although phenomenologically similar in some respects to those found in "ordinary" autism, differed sharply in the marked improvement evident between 4 and 6 years of age and in the degree of social interest...The quasi-autistic pattern seemed to be associated with a prolonged experience of perceptual and experiential privation, with a lack of opportunity to develop attachment relationships, and with cognitive impairment."

This sad epoch demonstrates both the potential for severely abnormal rearing practices to influence brain regions that are affected by typical causes of autism, and the resilience of the brain in compensating and restoring once the individual is placed in a more normal environment. But it does not provide evidence for the postnatal genesis of autism.

The research picture regarding the causes for Autism Spectrum Disorder remains complex, although there is certainly a very strong genetic component. While there are some genes, such as CHD8, the mutation of which almost always cause autism in a very low percentage of cases42 most mutations seem to confer small increases in risk. Similarly, while some environmental factors, such as rubella infection or fetal exposure to valproic acid, have been highly associated with autism risk, the increase in risk associated with others, such as living close to a highway, is small. It is very likely that the answer to what causes autism will not reside solely in genetics or in environment but in a combination of the two. Whatever factors go into the mix, they most likely have their effect during fetal life: a person with autism is born with autism.

2

Next Generation House Call

By Jamie L. Adams, M.D., Christopher G. Tarolli, M.D., and E. Ray Dorsey, M.D.

Jamie Adams, M.D., is an assistant professor in the Department of Neurology, with a dual appointment in the Center for Human Experimental Therapeutics, at the University of Rochester Medical Center. Her current research focuses on wearable sensor devices and telemedicine, and technologies in patient care, clinical trials, and drug development.

Christopher Tarolli, M.D., is an instructor of neurology and fellow in movement disorders and experimental therapeutics at the University of Rochester Medical Center. Tarolli received his medical degree from the State University of New York Downstate Medical Center's College of Medicine. Tarolli has participated in research evaluating the use of telemedicine and technology in individuals with Parkinson disease.

Ray Dorsey, M.D., M.B.A., is the David M. Levy Professor of Neurology and director of CHET, a center at the University of Rochester Medical Center that seeks to advance knowledge and improve health. Dorsey previously directed the movement disorders division and neurology telemedicine at Johns Hopkins. In 2015 he was recognized as a White House "Champion for Change" for Parkinson's disease.

Editor's Note: Just as online shopping is supplanting visits to the mall, and distance learning is part of the new wave in higher education, so is health care coming to a computer or mobile device near you. In the next few years, telehealth will increasingly become part of psychiatric and neurological care. Still to overcome is an unwieldy health care system that will need to adapt to practices that have the potential to lower costs and improve care.

NOT LONG AGO, doctors routinely made house calls. In fact, 40 percent of patient-physician encounters took place in the home as recently as the 1930s.[1] But with the advent of the automobile and the development of new diagnostic testing (e.g., x-rays, ECGs), care transitioned to clinics and hospitals. Today—in a gradual paradigm shift—broadband connectivity and point-of-care testing (e.g., glucometers) are fueling the rise of virtual visits.[2,3] By the 2030s, this next generation house call could be the dominant means of providing care to patients (Figure 1).[4–6]

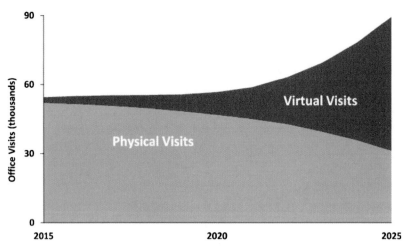

Figure 1. Projected number of in-person and virtual psychiatry and neurology office visits, 2015–25.

Telehealth in Psychiatry and Neurology

Telehealth is the use of telecommunications technology to provide health care at a distance. This includes care provision ranging from telephone calls and email to the use of web-based video conferencing technology (akin to Skype) via smart phone, tablet, or computer to virtually connect with a provider. It was initially implemented to increase access to care for individuals with acute conditions (e.g., trauma, stroke) in clinical settings, such as hospitals. Today's telehealth is focused on providing convenient care in the home to individuals with episodic conditions, such as rash or headache. In the future, the goal will be to lower the cost of care for individuals with chronic conditions anywhere, through personal computers, mobile devices, and email. (Figure 2).[7]

For different reasons, psychiatry and neurology have been early adopters of telehealth. The psychiatric evaluation relies less on the traditional physical exam than on a directed patient interview. This, in combination with the geographic mismatch between the supply of psychiatrists and the demand for psychiatric care, makes telehealth an appealing option. For over a generation, mental health professionals have used telehealth to deliver care to individuals in rural and urban locations, in clinics and in prisons.[8–10] Numerous high quality studies demonstrate the benefit of such interventions for diverse psychiatric conditions, including anxiety, depression, and post-traumatic stress disorder.[11–15]

Perhaps, more than any organization in the US, the Department of Veterans Affairs (VA) has embraced telehealth, especially for mental health care. In 2014, the VA had over two million telehealth visits, and mental health visits were among the most common. Most occur in community-based outpatient clinics in small communities that serve veterans, many of whom have significant mental health needs.[16–18] In addition to the VA, several start-ups have developed innovative health care models that are addressing substantial unmet needs among those with mental illness. Both the VA and these start-ups have started moving some of their care directly into patients' homes.

Teleneurology began in earnest in 1999 with a landmark paper by Drs.

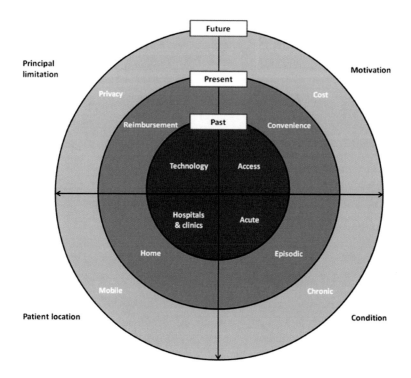

Figure 2. Past, present, and future of telehealth.

Steven Levine and Mark Gorman introducing the concept of "telestroke."[19] Motivated by the new clot-busting drug, tissue plasminogen activator (tPA), which has to be administered within hours of symptom onset, the idea was to enable timely, round-the-clock evaluation of individuals with suspected stroke by connecting stroke neurologists by video to local emergency rooms. Telestroke has improved stroke outcomes by increasing access to a time-sensitive medication, and studies have shown results comparable to in-person treatment.[20] In addition, telestroke has proven to be cost-effective, particularly since the benefits of improved acute stroke care include reduced long term health costs.[21] The idea has rapidly gained traction and spread around the country and the world.[22,23] Today, the largest single provider of acute stroke care is not a major medical center but a telehealth company, Specialists on Call, that cares for ten times as many individuals with stroke as any large stroke center.[24]

While telehealth for acute stroke has spread rapidly, its diffusion to chronic neurological conditions has been slow.[25] A 2012 survey of top neurology programs in the US found that the use of telehealth in conditions other than stroke is still very much in its infancy, and randomized controlled trials of such applications have been few and small.[26–28]

Move to the Home

In many ways, current care for individuals with psychiatric and neurological conditions could not be designed worse. In autism, we ask children with impaired social skills to travel to foreign environments and interact with multiple strangers to receive care. We expect older individuals with Parkinson's disease, whose mobility, cognition, and driving ability are compromised, to be transported by overburdened caregivers to large, complex urban medical centers.[29]

The need for patient-centered care is increasing. Neuropsychiatric conditions are now the leading cause of disability in the US and the third leading cause in the world.[30,31] The prevalence of autism spectrum disorders among American children is nearly two percent; the number of Americans with Alzheimer's disease is projected to reach 7.7 million by 2030, and the number with Parkinson's disease will almost double over the next generation.[32–34] In addition to those directly affected by such diseases, 40 million caregivers now help support adults with neuropsychiatric and other chronic medical conditions, including cancer, heart disease, and diabetes; telehealth has the potential to also increase their access to care and medical services.[35]

Beyond simply connecting patients to physicians, the telehealth model provides a platform for creating a patient-centered medical environment in the home. Rather than trying to coordinate the schedules of multiple providers or, worse, asking a patient to do so, clinicians—from psychologists to therapists—can connect to patients based on their mutual availability, all without the need for transportation. By delivering patient-centered care to individuals with chronic neuropsychiatric disorders directly into the home, telehealth can help reduce caregiver burden.

While home video visits for episodic conditions (e.g., sinusitis) are widely available for about $40 per visit, such access for chronic neurological and psychiatric disorders is still developing. Again, the VA is leading the way by studying and providing home telehealth care for depression, post-traumatic stress disorder, and Parkinson's disease.[13,28,36]

Although still a foreign concept to most, the latent demand for at-home care from a psychiatrist or neurologist is likely high. As part of a recent national randomized controlled trial of "virtual house calls" (video visits with a remote specialist in a patient's home), over 11,000 individuals from every state and 80 countries visited the study's one-page website. Of these, nearly 1,000 individuals with Parkinson's disease expressed interest in participating in the study.[37] Participant satisfaction, as in nearly all telehealth studies, was high.[38] In addition, these visits were shown to save patients and their caregivers three hours of time and 100 miles of travel per visit.[27]

Limitations and Barriers

Physical, policy, and social barriers are preventing the next generation house call from taking root. One clear limitation of telehealth is the physical exam. Most psychiatric disorders can be diagnosed via telehealth through a detailed patient interview, with limited need for a dedicated physical exam; telepsychiatry is demonstrated to be as effective as in-person visits for the diagnosis of conditions ranging from generalized anxiety disorder to schizophrenia.[39] In contrast, the exam is essential for the diagnosis of such neurological disorders as multiple sclerosis and myasthenia gravis. Telehealth for these and other disorders may serve as a complement in the ongoing care of individuals who already have a diagnosis confirmed during an in-person visit.[40] For example, having an individual with known amyotrophic lateral sclerosis, who has impaired mobility and compromised respiratory function, come repeatedly to a major medical center for ongoing care is not only illogical but also potentially dangerous. Breakdown of an effective doctor-patient relationship due to the loss of face-to-face visits is also cited as a potential pitfall of telehealth.[41] However, even in hospice care, a specialty where a strong doctor-patient relationship is paramount, telehealth has demonstrated benefit and been rated positively by clinicians, caregivers, and patients.[42]

Policy barriers to telehealth are hindering its development. The two largest such obstacles are reimbursement and licensure. While the VA— an integrated financing and delivery health system— has widely adopted telehealth, other insurers have been slower to adopt. Forty-eight Medicaid programs now cover telehealth, but coverage is varied and coverage in the home is frequently limited. At least 30 states now mandate that private insurers cover telehealth to the extent they cover in-person care, but again, care delivered into the home is often excluded.[43] The real laggard is Medicare, which in 2015 spent less than 0.01 percent of its budget on telehealth.[44] Medicare only covers telehealth in areas of health professional shortage, and only when delivered into clinical settings (e.g., medical offices, hospitals), which greatly limits access in a program whose fundamental purpose is to guarantee access.[45]

In addition to reimbursement barriers, state laws generally allow patients to receive care only from clinicians licensed in the state where the patient is located. Effective January 1, 2017, an interstate medical licensure compact took effect that should facilitate cross-state licensure for physicians in 18 (primarily western) states, but its impact is still uncertain.[46] State licensing boards also have variable policies on what activities (e.g., prescribing of medications) can be done remotely.[47]

The greatest barrier to adoption may be social. The fundamental purpose of telehealth is increasing access to care, but those who are least served and have the greatest need currently also have the least access to the internet and other technology necessary to take advantage of telehealth. These include individuals who are older, live in rural areas, and have lower incomes, less education, or more chronic conditions.[48–50] Overcoming this barrier will mean increasing access to broadband communications and the necessary technology, improving "tech-literacy," and providing support for those on the far side of the "digital divide."

Future

In *Singularity*, Ray Kurzweil, the chief engineer at Google and a futurist, posits the "law of accelerating returns."[51] He writes, "[Technological] change advances (at least) exponentially, not linearly … [and that] as a particular evolutionary process becomes more effective, greater resources are deployed toward the further progress of that process." In medicine, exponential advances in imaging and genetics have profoundly altered and advanced our understanding of neuropsychiatric conditions. Similarly, telehealth will fundamentally alter and advance the way we care for individuals with chronic neuropsychiatric conditions.

Where barriers have been addressed, such as at the VA and in Canada, adoption of telehealth has expanded exponentially. From 2005 to 2013 the number of telehealth visits at a VA medical center in Vermont, which began with mental health, has increased 20-fold.[18] From 2009 to 2014 the number of visits in the Ontario Telehealth Network has increased 10-fold.[52] While initial applications were primarily rural, urban use of telehealth is now dominant (Figure 3).

As telehealth brings care into the homes of individuals with psychiatric and neurological conditions, three changes are likely to occur. First, the use of telehealth will rise exponentially. The exact timing and rate of adoption will be determined by policy changes, especially in Medicare's coverage of telehealth. However, even absent such change, adoption will occur in other settings, either as a result of economic incentives for the use of telehealth, or simply by immense social forces (e.g., the mobility of the nuclear family, the aging of the population, geographically dispersed caregivers, broad adoption of internet for other services) driving demand.[7,53]

Second, the aggregate number of patient visits will increase. Few patients feel they are receiving too much care, and empirical evidence suggests when barriers to access are reduced, utilization increases. Last year Kaiser Permanente in Northern California had more virtual (phone, email, and video) than in-person visits, which have remained relatively stable.[54] The US Congressional Budget Office is concerned that by increasing visits, telehealth will lead to higher costs.[55] However, those concerns are short-sighted

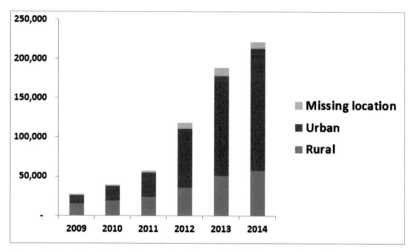

Figure 3. Number of telehealth visits by Ontario Telehealth Network, 2009-14.
Source: O'Gorman LD, et al. Telehealth and e-Health 2016;22:473-952

and misplaced. Large, high-volume, and centralized health systems that are most at risk for increased utilization, such as in Canada or at the VA and Kaiser, have been the largest adopters of telehealth; they realize that such patient-centered care is far less expensive than institution-delivered care. Indeed, in Parkinson's disease, more visits to a neurologist are associated with fewer hospitalizations and lower overall Medicare expenditures; telehealth offers a mechanism to facilitate and bear this increase in specialty neurological care through improving access, while limiting the economic burden.[56]

Third, the number of in-person office visits will slowly decline. Currently, over 50,000 psychiatry and neurology office visits occur annually in the US.[57] This is unlikely to change in the near term, but virtual visits in the home will eventually replace some office visits, which are costly to patients in terms of time and travel (the average 20-minute doctor's visit takes two hours of time and travel) and to clinics in terms of labor, space, waiting rooms, and parking lots.[58] The future implications for this change will be profound, just as they have been in the retail sector, where stores from Walmart to Macy's are closing, shopping malls are disappearing, and home delivery volumes are rising.[59–62] For psychiatry and neurology clinics, space and labor needs will decrease and beyond the increase in virtual visits, demand for traditional in-person home visits (e.g., for support, relationship development) will likely increase as care convenience becomes a priority. Mutually beneficial relationships with local clinicians will need to be cultivated to ensure proper oversight of care and to address emergencies. For clinicians, training for digital medicine will have to begin.

For now, as the burden of neuropsychiatric conditions rises along with the demand for convenient, patient-centered care, telehealth is poised to deliver care where it has always been needed most—at home.

3

The Four Pillars of Alzheimer's Prevention

By Dharma Singh Khalsa, M.D., and George Perry, Ph.D.

Dharma Singh Khalsa, M.D., is the president/medical director of the Alzheimer's Research and Prevention Foundation (AARPF) and the author of *Brain Longevity* (Warner Books, 1997). He is also a clinical associate professor, Division of General Internal Medicine, Geriatrics, and Integrative Medicine at the University of New Mexico Health Sciences Center in Albuquerque, and an associate editor of *The Journal of Alzheimer's Disease*. Born in Ohio and raised in Florida, Khalsa graduated from Creighton University School of Medicine in 1975 and received his postgraduate training in anesthesiology at the University of California, San Francisco. He is board certified in anesthesiology and pain management and a diplomat of the American Academy of Anti-Aging Medicine.

George Perry, Ph.D., is dean of the College of Sciences and holds the Semmes Foundation Distinguished University Chair in Neurobiology at the University of Texas at San Antonio. He obtained his Ph.D. from Scripps Institution of Oceanography in 1979 and received a postdoctoral fellowship in the Department of Cell Biology at Baylor College of Medicine. Perry is editor-in-chief for the *Journal of Alzheimer's Disease* and is a foreign correspondent member of the Spanish Royal Academy of Sciences, the Academy of Science Lisbon, and a foreign member of the Mexican National Academy of Sciences. He is a recipient of the National Plaque of Honor from the Republic of Panama Ministry of Science and Technology. Perry's research is primarily focused on how Alzheimer's disease develops and the physiological consequences of the disease at a cellular level.

Editor's Note: Much is yet to be discovered about the precise biological changes that cause Alzheimer's, disease, why it progresses more quickly in some than in others, and how the disease can be prevented, slowed, or stopped. And while researchers continue to search for the magic pill that can prevent or halt the spread of amyloid in the brain, our authors believe that changing or modifying one's lifestyle and attitude can make a difference in both prevention and treatment.

MEMORY PROBLEMS COME IN all shapes and sizes. Some people tend to forget where they put their cell phone, cannot easily recall names. can't recall taking their medication, or remember the birthday or anniversary of a loved one. Whether they admit to themselves that their forgetfulness seems to happen with greater frequency or they worry about losing their memory as they age, they are right to be concerned. Because our aging population is on the rise, Alzheimer's disease (AD)—an irreversible, progressive form of dementia that slowly destroys memory and thinking skills as people age and is ultimately fatal—has steadily risen from about four million in the late 1990s to 5.4 million today.[1]

The disease is currently ranked as the sixth leading cause of death in the US, but estimates by the National Institute on Aging indicate that it may rank third, just behind heart disease and cancer, as a cause of death for older people.[2] But here is some good news: Whether you want to reverse cognitive deficits now or avoid them later, more and more studies are suggesting that there is much you can do to keep your mind sharp.

While a pharmaceutical approach to preventing AD has proved elusive, practical lifestyle choices to reduce AD are based on good science and good sense. The secret may lie in *epigenetics*, the effect one's lifestyle has on one's genes, and thus on the risk for disease. Of course, the wisdom that lifestyle has an impact on health is not new; we have been reciting adages such as "an apple a day keeps the doctor away" for ages. Research in a variety of areas has confirmed that sensible everyday choices can significantly reduce the risk of AD. According to the National Institutes of Health, $991 million was dedicated to AD research in 2016, but how much of that went towards lifestyle-modification and prevention is unclear.[3]

Funding uncertainty notwithstanding, the positive effects of a healthier lifestyle on cognition were recently documented for the first time in a longitudinal study. The two-year, 1,200 participant *Finnish Interventional Geriatric Study for the Prevention of Cognitive Disability (FINGER)* showed that a healthy diet, exercise, socialization, and mental stimulation can dramatically reduce the development of AD in people at risk for cognitive decline.[4] The French *MAPT Study: A Multidomain Approach for Preventing Alzheimer's Disease* also suggested that lifestyle modification has an effect in reducing risk factors.[5] This multi-domain approach is consistent with the four-pillar strategy recommended by a number of reputable sources, including the Alzheimer's Research and Prevention Foundation (ARPF), the Dana Alliance, the American Association of Retired Persons, and the Alzheimer's Association.

The aforementioned studies add substantially to mounting scientific evidence that suggests lifestyle and psychological well-being play a critically important role in Alzheimer's prevention. We have taken them into account, along with our own findings, in fine-tuning our longstanding recommendations for staving off and even helping to reverse AD to the following four strategies. The secret to AD prevention is tied to maintaining connections: between your brain cells, other people, and your well-being.

Pillar 1: Diet and Supplements

Diet is one of the most important targets for lifestyle modification to prevent AD. Many people still blindly follow the Standard American Diet, or SAD. According to the US Government, about 75 percent of all Americans do not consume an adequate amount of vegetables and fruits, while most exceed the recommended amount of sugars, saturated fats, sodium, and calories. Studies show that rejecting SAD may be critical in the fight against AD.[6]

The science reveals that those who eschew processed foods and choose whole, real-food options have the least decline in mental faculty. Research published in the Alzheimer's Association's journal *Alzheimer's & Dementia,* for example, confirms that making the switch from a fat- and meat-heavy way of eating to a primarily plant-based diet—no matter how old a person is when doing so—can slow and possibly reverse memory loss. The components of a healthy diet may enhance cognitive performance by one or more of several actions: affecting synaptic plasticity, synaptic membrane fluidity, glucose utilization, mitochondrial function, or reducing oxidative stress.[7]

Many studies highlight the Mediterranean diet that is rich in vegetables, fruit, nuts, olive oil, and fish or seafood. Researchers at UCLA discovered that study participants who followed this eating plan, which is modeled on the traditional diet of certain Mediterranean peoples, had lower levels of AD's hallmark amyloid-beta plaques in the spaces between their brain nerve cells, along with fewer telltale tangles of tau protein—meaning those important cell connections were firing properly.[8] And at the Mayo Clinic, through MRI scans, researchers found that participants who followed the Mediterranean diet for a year had greater thickness in parts of their brain's cortex that play a role in memory. Those on the SAD diet, on the other hand, lost cortex. These findings have implications for maintaining cognitive function: positive associations of the Mediterranean Diet scores were observed with average cortical thickness in parietal and frontal lobes, and in regions of the brain that mediate or support elements such as memory, executive function, and language.[9] Americanized versions of the Mediterranean diet, as well as the MIND (Mediterranean-DASH Intervention for Neurodegenerative Delay) and DASH (Dietary Approaches to Stop Hypertension) diets, have also shown promising results. Research from Rush University, where the MIND diet was developed by nutritional epidemiologist Martha Clare Morris, revealed that the MIND diet could turn back your mental aging clock the equivalent of up to 7.5 years. Although this is now widely accepted by researchers, further confirmative studies are ongoing.[10]

The ARPF nutrition plan has much in common with both the Mediterranean and MIND diets. Some of the organization's main tenets are:

- *Vegetarian foods:* A vegetarian diet—full of fruits and vegetables, nuts and seeds, legumes and soy—improves focus and begets higher productivity. Wild-caught salmon is the only animal protein the Alzheimer's Research & Prevention Foundation's diet recommends for its brain-friendly omega-3 fats, advising eating it only two to three times a week.

- *Juicing:* Fresh juices are alive with the vitamins, minerals, trace elements, and phytonutrients needed to strengthen the brain.

- *Supplements:* Take a high-potency multivitamin and multi-mineral supplement that includes folic acid. Memory specific supplements of omega-3 oils, phosphatidyl-serine, coenzyme Q10, alpha lipoic acid, huperzine-A, and resveratrol are also recommended.

As previously noted, we suspect that certain genes can influence risk of developing AD. But well-chosen foods and their nutrients may move gene expression toward a sharp brain. "Genetics are not our destiny," says Victor S. Sierpina, M.D., professor of family and integrative medicine at the University of Texas Medical Branch in Galveston. "How we eat can have a major impact in reducing our risk of developing this feared condition." By moving away from the SAD diet to a more Mediterranean-type diet, it is possible to eat for optimal brain health.

Pillar 2: Physical and Mental Exercise

The evidence is convincing: Both physical and mental exercise are absolutely essential in preventing AD. Exercise increases blood flow to the brain, augments crucial brain compounds such as brain-derived neurotrophic factor (BDNF), and, perhaps most significantly, causes neurogenesis, or the growth of new brain cells. In a study at Columbia University, researchers showed that older men who exercised on a treadmill four times a week for 30 minutes grew new cells in their dentate gyrus, an important area of the brain related to memory and cognition such as executive function.[11] And guess what? One can experience these brain-boosting effects of exercise regardless of one's age or existing level of fitness or cognitive decline.

Current wisdom recommends 150 minutes a week of cardio (aka aerobic) exercise, plus several sessions of strength training. But the benefits of even mild exercise begin to accrue right away. Just getting out and taking a 20 to 30- minute brisk walk three times a week will improve brain and memory function. Like diet, exercise also creates a healthy epigenetic response. Those who are already in good physical condition should add more variety and intensity to their workouts. Get a trainer, join a gym, play tennis, swim, or take a boot camp, Zumba, or cycling class. Find enjoyable activities and make them part of your routine.

Additionally, keeping one's mind active is an important aspect of AD prevention. There are a variety of ways to do this. One of them, reading, is one of the best ways to stay sharp—not only does learning take place, but the mind is forced to think and engage outside of everyday tasks. Other simple strategies—or what are sometimes called brain-aerobic activities—include playing and listening to music, creating and viewing art, or completing crossword puzzles. All stimulate and challenge the brain, giving it a nice "workout." Remember, it's not just about physical fitness, it's about mental conditioning as well.

Pillar 3: Yoga/Meditation

Chronic stress is a major risk factor for AD.[12] It may be useful to experience stress if one is running for his or her life, but not when just trying to live one's life. Stress has a detrimental effect on genes, causing them to express themselves in unhealthy ways, such as by producing inflammation, a trademark of AD. The frenzied pace of life that people experience in today's world is only accelerating, so it is helpful to find a regular activity to soothe the harmful force of stress on the brain.

Published research over the past 13 years reveals that a simple, 12-minute yoga/meditation technique called Kirtan Kriya (KK) has significant brain boosting benefits. KK has been examined at leading medical schools, with the impressive, perhaps surprising, results published in more than one medical journal, including the *Journal of Alzheimer's Disease*.[13]

The actual age of KK is unknown. It was passed down from master to student for generations in the East until Yogi Bhajan (1929-2004) brought it to the West in about 1970. *Kirtan* means "singing" and *Kriya* means "an action with specific effects." KK involves singing the sounds *Saa Taa Naa*

Maa (a mantra) while repeating sequential movements (mudras) with the fingertips.

Ancient yogis did not have imaging or blood tests to unravel the biochemical changes created by KK and other yoga exercises, but modern science has shown that practicing KK reduces stress levels and increases blood flow to parts of the brain that are central to memory and brain function.[14] For example, KK activates the anterior cingulate gyrus (ACG), an important brain region for stress balance and emotional and cognitive control. A robust ACG is essential to memory: Research in the elderly who've maintained sharp minds shows they have preserved their ACGs and other significant brain areas as they have aged.[15]

The prefrontal cortex (PFC), the chief executive officer of your brain, essential for planning and organization, is also activated by meditation. So is the posterior cingulate gyrus (PC), one of the first areas to decline in function when memory loss strikes. Such findings have led Andrew Newberg, M.D., of Thomas Jefferson Medical School, to say, "There is a true anti-aging effect in long-term practitioners of KK; they have bigger brains."

A study at the University of Pennsylvania, which followed people with early cognitive decline for eight weeks, demonstrated that practicing the yoga/meditation technique started reversing memory loss and reduced anxiety, two hallmarks of early AD.[16] A UCLA study of family dementia care-

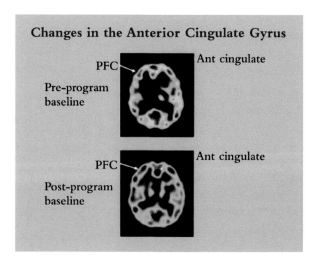

Changes in the Anterior Cingulate Gyrus

PFC

Ant cingulate

Pre-program baseline

PFC

Ant cingulate

Post-program baseline

Figure 1. Increased Size of PFC and ACG after Eight Week Program of 12 Minutes a day of KK.[13] (Courtesy of the Alzheimer's Research and Prevention Foundation.)

givers revealed that KK not only lowered their stress and improved their memory, but also reduced inflammatory genes and increased the enzyme telomerase by 43 percent, the largest increase ever recorded. Increasing this enzyme elongates the DNA protective cap, the telomere, which is crucial for a long life and a sharp mind.[17]

Additionally, at West Virginia University, subjects with the earliest form of memory loss, subjective cognitive decline (SCD), showed an improvement in cognitive function with KK.[18] And a landmark study at UCLA found that subjects with an advanced form of early memory loss called mild cognitive impairment (MCI) had better memory outcomes with KK than those who practiced a standard memory-improvement approach.[19] KK apparently enhanced brain cell connectivity as well. Importantly, the positive benefits lasted through the six-month follow up period of the study.[20]

KK has practical advantages. It only takes 12 minutes a day and requires no equipment or lengthy, expensive training sessions. One can practice KK at home with an easy-to-follow CD, for example, and it is completely safe, with no side effects reported. Its lack of time requirements makes the practice perfect for caregivers, and it's easy for seniors with decreased mobility and activity levels.

• Reduces Stress	• Decreases Anxiety and Depression
• Improves Sleep	• Improves Psychological Well-being
• Reverses Memory Loss	• Enhances Neurotransmitter Function
• Increases Energy Levels	• Increases Blood Flow to Significant Brain Areas
• Upregulates Positive Genes	• Reduces Some Risk Factors for Alzheimer's Disease
• Increases Telomerase By 43%	• Down Regulates the Genes that Cause Inflammation

Figure 2. A summary of the effects of KK. (Courtesy of the Alzheimer's Research and Prevention Foundation.)

Pillar 4: Psychological Well-Being

Meditation also enhances psychological well-being (PWB) by promoting acceptance of self and others, increasing self-confidence, reducing negativity, and providing a foundation for independent living, sustained personal growth, socializing with like-minded people, service to others, and aging with purpose. These PWB factors lower the risk for cognitive decline and help reduce cholesterol and inflammation.[21] In fact, Purpose in Life is a new movement in neuroscience that links the belief that one's life has meaning and purpose to a robust and persistently improved physiological health outcome—not only to treat AD, but also to treat spinal cord injuries, stroke, and immunological and cardiovascular issues that include but extend beyond the brain.[22]

Positive emotions—love, compassion, and appreciation—counteract the physiology of the stress response and support a healthy brain throughout life. Beyond that, PWB may create an enhanced sense of spirituality, which preliminary studies suggest slows the progression of AD.[23] Moreover, per Helen Lavretsky, M.D., a geriatric psychiatrist at UCLA, spirituality is a way to develop personalized, patient-centered healthcare. There is evidence of a close relationship between spirituality, cognitive health, and successful aging.[24] Finally, in a very recent and, as of yet, unpublished three-year study, spirituality was associated with lowered atrophy rates in brain regions related to memory, visuospatial attention, and behavioral deficits in subjects at risk for AD.[25]

As it currently stands, or until the pharmaceutical world can meet the enormous challenge of discovering the magic anecdote that can make amyloid disappear, living a healthy life offers the best chance for aging AD-free and nourishing a sharp mind. Small, easily achieved shifts in one's daily routine can make all the difference in brain health. If everyone made such shifts, it is likely that the widespread prediction of a continuing Alzheimer's epidemic would shift, too, with fewer reported cases.

4

Gut Feelings on Parkinson's and Depression

By Timothy G. Dinan, M.D., Ph.D., and John F. Cryan, Ph.D.

Ted Dinan, M.D., Ph.D., is professor of psychiatry and a principal investigator in the APC Microbiome Institute at University College Cork in Ireland. He was previously chair of clinical neurosciences and professor of psychological medicine at St. Bartholomew's Hospital in London. Prior to that, he was a senior lecturer in psychiatry at Trinity College Dublin. Dinan has worked in research laboratories on both sides of the Atlantic and has a Ph.D. in pharmacology from the University of London. He is a fellow of the Royal Colleges of Physicians and Psychiatrists and a Fellow of the American College of Physicians. In 1995, Dinan was awarded the Melvin Ramsey Prize for research into the biology of stress. His current research is funded by Science Foundation Ireland, the Health Research Board and European Union FP7. He has published over 400 papers and numerous books on pharmacology and neurobiology and is on the editorial boards of several journals.

John F. Cryan, Ph.D., is professor and chair in the Department of Anatomy and Neuroscience, University College Cork in Ireland. He is also a principal investigator in the APC Microbiome Institute. Cryan has published over 330 peer-reviewed articles, is a senior editor of *Neuropharmacology* and of *Nutritional Neuroscience*, and is on the editorial board of 15 other journals. He has edited three books, including *Microbial Endocrinology: The Microbiota-Gut-Brain Axis in Health and Disease* (Springer Press, 2014). His awards include UCC Researcher of the Year in 2012 and the University of Utrecht Award for Excellence in Pharmaceutical Research in 2013. Cryan was named to the Thomson Reuters Highly Cited Researcher list in 2014, was a TEDMED speaker in Washington in 2014, and was named president-elect of the European Behavioural Pharmacology Society in 2015.

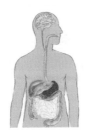

Editor's Note: The gut-brain axis is one of the new frontiers of neuroscience. Microbiota (the collective bacteria, viruses, fungi and other microorganisms that live in the digestive tract), sometimes referred to as the "second genome" or the "second brain," may influence our health in ways that scientists are just now beginning to understand. Scientists now believe that the microbiota and all that it involves may be a way to treat any number of disorders, including Parkinson's disease and depression.

OVER THE PAST FEW YEARS, the gut microbiota has been implicated in developmental disorders such as schizophrenia and autism, neurodegenerative disorders such as Alzheimer's disease and Parkinson's disease, mood disorders such as depression, and even addiction disorders. It now seems strange that for so many decades we viewed the gut microbiota as bacteria that did us no harm but were of little benefit. This erroneous view has been radically transformed into the belief that the gut microbiota is, in effect, a virtual organ of immense importance.

What we've learned is that what is commonly referred to as "the brain–gut–microbiota axis" is a bidirectional system that enables gut microbes to communicate with the brain and the brain to communicate back to the gut. It may be hard to believe that the microbes in the gut collectively weigh around three pounds—the approximate weight of the adult human brain—and contain ten times the number of cells in our bodies and over 100 times as many genes as our genome.[1] If the essential microbial genes were to be incorporated into our genomes, it is likely that our cells would not be large enough for the extra DNA.

Many of those genes in our microbiota are important for brain development and function; they enable gut bacteria to synthesize numerous neurotransmitters and neuromodulators such as γ-aminobutyric acid (GABA), serotonin, dopamine, and short-chain fatty acids. While some of these compounds act locally in the gut, many products of the microbiota are transported widely and are necessary for the proper functioning of diverse organs. This is a two-way interaction: gut microbes are dependent on us for their nourishment. Any pathological process that reduces or increases food intake has implications for our microbes.

The routes of communication between gut microbes and the brain are not fully known, but they most likely include neural, endocrine, immune, and metabolic pathways.[5] In particular, the long, meandering vagus nerve is thought to be a key bidirectional information pathway. Multiple chemicals are produced in the gut and travel to the brain through the blood stream. Many of these chemicals—tryptophan, leptin, and ghrelin—are produced by or have their production controlled by gut microbes and are involved in regulating mood and appetite. Cytokines, which are key immune molecules produced at the level of the gut, can travel through the blood stream and influence brain function in certain regions. One such region is the hypothalamus, where there is a deficient blood brain barrier, which in other brain regions plays a vital protective role by blocking substances in the body from entering the brain. It is certain that because of this structural difference with most other brain parts, the hypothalamus can be influenced by molecules in the blood which do not gain access to the brain elsewhere.

Linking Parkinson's and Microbiota

Symptoms of Parkinson's disease (PD)—the "shaking palsy"—may have been first described by the Greek physician Galen who worked in ancient Rome. But the disease is named for James Parkinson of Shoreditch in London, who provided the first detailed description of the condition in 1817.[2] This debilitating neurological disorder sometimes strikes young people but is more common in seniors. Its motor symptoms–slow movements, resting tremor, rigidity, and postural instability—predominate and can have an impact on every aspect of daily activities.

It was over a century and a half after James Parkinson that the underlying pathophysiology of the condition was understood to involve neurons in the substantia nigra, a part of the brain that plays a pivotal role in reward and movement.[3] These neurons synthesise and release the neurotransmitter dopamine, which is the focus of current treatment. Dopamine enhancement does not control symptoms consistently, however. Many patients with PD need improved supplemental therapies. PD patients frequently have non-motor symptoms that may include loss of smell (olfaction), gastrointestinal problems (especially constipation), cardiovascular dysfunction, and urogenital system difficulties. At least 50 percent develop depression during the course of their illness and approximately one in three are depressed at any point in time, sometimes prior to the onset of overt motor symptoms.

It is conceivable that PD occurs because of toxins produced by the gut microbiota or because of a failure to produce key essential neuronal dopamine specific nutrients, which are required by dopamine producing cells. The microbiota that we initially develop is determined by the mode of birth delivery of the infant (vaginally or by caesarean section), diet, and exposure to antibiotics.[4] It has been shown that the core microbiota of aged individuals is characteristically distinct from that of younger adults and that age-related shifts in its composition and diversity are linked to adverse health effects in the elderly. Of interest is the fact that the number of probiotic, health-benefitting *Bifidobacteria* in the intestines decreases with age. In contrast, healthy young babies have large numbers of *Bifidobacteria*. Taking probiotic *Bifidobacteria* does not entirely replace these decreasing bacteria as probiotics are tourists in the intestine and do not result in permanent colonization. However, it seems likely that maintaining a healthy and diverse diet, exercising, and avoiding obesity as we age can have a beneficial impact on such microbes.

Linking Parkinson's and the Vagus Nerve

In the development of PD, aggregation of the protein alpha-synuclein plays such an important role that many researchers consider it central to the disease process.[6] Some believe that the protein—which is manufactured by microbes—originates in the gut and spreads up the vagus nerve like a prion (a small infectious molecule formed by abnormally folded protein). Dopamine-containing neurons within the substantia nigra—the brain region apparently responsible for the motor symptoms of PD—are especially sensitive and vulnerable to the accumulation of alpha-synuclein. The protein may also play a role in the autonomic disorders of the cardiovascular and gastrointestinal systems that are seen in PD.

The risk of developing PD is reduced in individuals who have undergone vagotomy, a surgical procedure to cut the vagus nerve, which was commonly used to treat peptic ulcer disease prior to the introduction of effective medications such as *H2 antagonists* and proton pump inhibitors in the 1970s and 80s. A Danish study[7] investigated the risk of PD in patients who underwent vagotomy and hypothesized that truncal vagotomy (which severs the entire nerve) is associated with a protective effect, whereas super-selective vagotomy (which only partially severs the nerve) has only a minimal effect. Their findings confirmed their hypothesis: When compared to the general population, risk of PD was decreased after truncal vagotomy. These epidemiological findings, which support the view that PD initially commences in the gut and not the brain, strongly implicates the vagus nerve in the disorder's development.

Other studies provide a mechanistic explanation. Researchers have traced the upward spread of alpha-synuclein from the gut along vagal nerve fibres.[6] This suggests a neuronal route for the transport of probable PD pathogens or toxins from the enteric nervous system in the gut to the brain, rather than via the bloodstream.

The Role of Gut Dysbiosis

The gut microbiome in PD has only recently been investigated, and researchers speculate whether microbes are altered in PD. Sheperjans and colleagues[8] compared the fecal microbiomes of 72 PD patients and 72 control subjects by sequencing the bacterial 16S ribosomal RNA gene, which is used to identify specific bacteria. Associations between clinical parameters and microbiota were analysed, considering potential confounding factors. On average, the abundance of Prevotellaceae microbes (normally found in very high concentrations in the gut of vegetarians) in faeces of PD patients was reduced by 77.6 percent, compared with controls. Relative abundance of Prevotellaceae of 6.5 percent or less had 86.1 percent sensitivity and 38.9

percent specificity for PD. An analysis based on the abundance of four bacterial families and the severity of constipation identified PD patients with 66.7 percent sensitivity and 90.3 percent specificity.

Interestingly, the relative abundance of a different type of microbe, Enterobacteriaceae, was positively associated with the severity of postural instability and gait difficulty. Enterobacteriaceae can be harmless or pathogenic as in the case of salmonella. These findings suggest that the intestinal microbiome is altered in PD and is related to the motor phenotype, though one cannot conclusively dismiss the possibility that these changes are epiphenomenal. Such findings raise the exciting possibility that changes in the gut microbiota may be used as a diagnostic marker, though further large-scale studies are required.

Mazmanian's group[9] have investigated the genesis of PD features in mice genetically engineered to overexpress alpha-synuclein, which go on to develop features of PD. These mice are now one of the most widely used animal models in studying the disorder. Intriguingly, when they are raised germ-free (i.e., without gut microbes), their tendency to develop motor abnormalities is significantly reduced. When they are given a combination of short chain fatty acids, on the other hand, they show microglial activation in the brain and aggregation of alpha-synuclein with onset of motor features. These changes are inhibited by treatment with the antibiotic minocycline, which acts on a broad range of bacteria. If germ free alpha-synuclein overexpressing animals are given a humanized microbiota from a patient with PD, the emergency of pathology is far greater than those transplanting with the microbiota from a healthy subject. These findings pinpoint several potential lines of treatment, including the use of short-chain fatty acid antagonists, antibiotics, and microbiota transplantation. Time will tell if any of these potential therapeutic options prove fruitful.

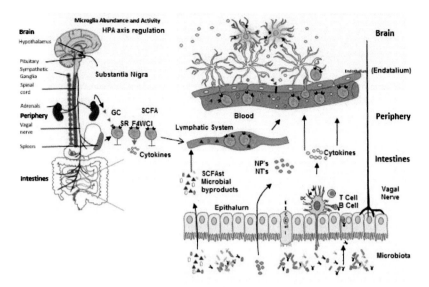

Figure 1. Communication within the microbiota-gut-brain axis involves the co-ordination of multiple factors, including vagal nerve activation, cytokine production, neuropeptide/neurotransmitter release and short chain fatty acid release. When these signals penetrate the blood brain barrier, they influence the maturation and activation state of the microglia, key immune cells in the brain. Once activated, microglia play a fundamental part in immune surveillance, synaptic pruning, and clearance of debris. The HPA axis by cortisol production can in turn suppress the activation state of brain microglia, as well as influence cytokine release and trafficking of monocytes from the periphery to the brain.

Linking Parkinson's and Depression

PD and depression frequently go hand in hand. Given the impact of PD symptoms on quality of life, it is not surprising that many patients become depressed. However, other pathophysiological aspects of PD may also render sufferers vulnerable. In a recent study, the status of non–dopamine neurons, including those that primarily express the neurotransmitters acetylcholine, norepinephrine, and serotonin, was examined in post-mortem brains of PD patients.[10] In comparison with controls, PD brains showed widespread deficits of both serotonin and serotonin transporter, especially in brain areas called the caudate nucleus and the middle frontal gyrus. Significantly reduced norepinephrine levels were found in all brain regions tested, while acetylcholine transporter levels were not altered. The non-dopaminergic changes in transmitters such as serotonin and noradrenaline might well account for the high prevalence of depression in PD. While the core movement problems seen in PD are induced by a lack of dopamine, the co-morbid mood related features are likely due to changes in other neurotransmitters.

Microbiota in Depression

Recently, depression per se has also been associated with significant microbiota changes. Jang et al[11] analysed faecal samples from 46 patients with major depression and 30 healthy controls. Those who were acutely depressed had higher levels of *Bacteroidetes*, *Proteobacteria*, and *Actinobacteria*, whereas level of *Firmicutes* was significantly reduced. The exact physiological implications of this are not fully understood. However, a negative correlation was observed between *Faecalibacterium* and the severity of depressive symptoms.

Another study, conducted at the APC Microbiome Institute,[12] found an association between depression and decreased gut microbiota richness and diversity. When faecal microbiota was transplanted from depressed patients into microbiota-depleted rats, it induced behavioral features characteristic of depression, including anhedonia and anxiety-like behaviors, as well as alterations in tryptophan metabolism. This suggests that the gut microbiota may play a causal role in the development of mood disorder.

Faecal Microbiota Transplantation (FMT) for Parkinson's?

The concept of FMT for treatment of human gastrointestinal disease was described almost 2,000 years ago by a Chinese physician, Ge Hong, who orally administered human faecal suspension to treat patients with food poisoning or severe diarrhoea. Borody et al[13] noted that FMT may have been first used in veterinary medicine by the Italian anatomist Fabricius Aquapendente in the 17th century and that the first report of FMT in modern times was in the late 1950s for the treatment of pseudomembranous colitis.

Over the past few years, FMT has been increasingly used to treat resistant *Clostridium difficile* infection. It is remarkably effective in this application. There are obviously challenges in determining appropriate donors and which route of administration works best. Some gastroenterologists use a nasogastric route (the insertion of a plastic tube through the nose, past the throat, and down into the stomach), while other use enemas. Capsules filled with faecal material and taken orally are another option.

While *Clostridium difficile* is the only condition for which FMT is currently used, others are being considered. There have been case reports suggesting efficacy in PD, but it is yet to be determined whether it is a realistic therapeutic option. If gut microbes are therapeutically beneficial, then probiotics or cocktails of probiotics (polybiotics) may also prove beneficial. We await studies of these approaches.

Psychobiotics

Probiotic bacteria with a mental health benefit more generally have been termed psychobiotics. Several bacteria have been profiled pre-clinically and show promise, but large scale translational studies are required. A recent study found a *Bifidobacterium longum* strain to reduce anxiety levels in healthy subjects and to reduce levels of the stress hormone cortisol.[14] It is tempting to speculate that psychobiotics will someday be used to treat mild forms of both depression and anxiety.

Looking Forward

In the next few years, new research will hopefully affirm the validity of modulating the gut microbiota as a viable therapeutic strategy for treating neurological and psychiatric disorders. For example, emerging evidence supports the view that dysbiosis within the gut is the trigger for degeneration of central dopamine neurons in PD. If this proves to be the case, therapies that actually target the disease process may quickly become available, providing radical, more effective alternatives for patients. Such a breakthrough is badly needed.

5

Genetics and ALS: Cause for Optimism

By Roland Pochet, Ph.D.

 **Roland Pochet**, Ph.D., Honorary Professor of Cell Biology at the Faculty of Medicine of Université Libre de Bruxelles, is a neuroscientist who studies stem cell transplantation for amyotrophic lateral sclerosis. Pochet spent two years in Israel with an Excellent in the Life Sciences (EMBO) fellowhip and in labs at the University of Texas and Vanderbilt University. From 2006 to 2013, he was chair of the European Cooperation in Science and Technology (COST), an international organization that includes 36 countries. He created the European Calcium Society, is general secretary of the Belgian Brain Council, and evaluates projects for the European Commission. Pochet is member of the European Dana Alliance for the Brain and International Scientific Committe of AriSLA (Italian SLA).

Editor's Note: While drug development has done little to slow the devastating symptoms of amyotrophic lateral sclerosis (ALS), there is some good news in the fact that scientists have identified some 100 related genes and believe that genetic research offers the best hope for treatments. More good news came on the heels of the Ice Bucket Challenge, which raised $220 million globally and has fueled renewed optimism and energy in the ALS community.

"I compare my nervous system to a plaster building. Every day, large crumbly pieces break away. For now, the building remains standing, but the moment of collapse seems inescapable."

— Frédéric Badré, a French artist and writer,
who died in 2016 at 50 years old from ALS

SINCE JEAN-MARTIN CHARCOT FIRST described amyotrophic lateral sclerosis (ALS) in 1869, the mechanism underlying the degeneration and death of motor neurons that characterize the disease has remained a mystery. Charcot, considered the Father of Neurology and a professor of pathological anatomy at the University of Paris, was also the first to distinguish Multiple Sclerosis, Parkinson's, and ALS as separate diseases that involve impaired or lost motor neurons affecting muscle movement. His prognosis for those with ALS was also astute: "Six months to a year after onset, the full complement of symptoms is present and more or less patently manifested. Death occurs on average about two to three years after the appearance of the bulbar symptoms."[1]

Beyond Charcot's core definition, ALS is now known to involve the spinal cord and dysfunction of motor pathways in the cortex. Per the ALS Association, "Once ALS starts, it almost always progresses, eventually taking away the ability to walk, dress, write, speak, swallow, and breathe and shortening the life span. How fast and in what order this occurs is very different from person to person. While the average survival time is three years, about

20 percent of people with ALS live five years, 10 percent will survive ten years, and five percent will live 20 years or more." ALS is diagnosed and confirmed in about 1 in 500 to 1 in 1,000 adult deaths; 500,000 people in the US will develop this disease in their lifetimes.[2]

The diagnoses of ALS in the New York Yankees first baseman Lou Gehrig—nicknamed "The Iron Horse" because of his remarkable streak of playing in 2,130 consecutive games—in 1939 brought national and international attention to the disease. After ALS took his life at the age of 37, the illness became known as "Lou Gehrig's disease." Other well-known people who are struggling with or have died from ALS include Mao Tse-tung, Stephen Hawking, Charles Mingus, and David Niven. Tuesdays with Morrie, the bestselling memoir about author Mitch Albom's conversations with his former sociology professor Morrie Schwartz, depicted the disease's devastating impact.

In 1962, the British neurologist Walter Russell Brain, introduced the concept of "motor neuron disease," a group of diseases of which ALS is the most common. In the sixth edition of his book Brain, published in 1962, he made it clear that the three syndromes: upper motor neuron degeneration, lower motor neuron degeneration, and mixed upper motor and lower motor neuron degeneration should be regarded as manifestations of the same underlying disorder.[3]

ALS has been artificially distinct from other motor neuron diseases in the sense that it occurs both sporadically (sALS) and within families (familial ALS, fALS). Although some feel this distinction is practical, others feel it is artificial, since ALS is now believed to be an umbrella term for many different forms of the motor neuron disorder (just as there are various types of flu).

Today, ALS diagnosis remains difficult because these diverse forms involve a spectrum of manifestations whose heterogeneity extends to the site of onset, the degree of upper and lower motor neuron involvement, the rate of motor progression, and the presence and severity of cognitive and behavioral symptoms.

The enormous progress made in both technology and genetics over the past 20 years has led to the discovery of the first genetic mutation that causes ALS and the identification of more than 100 ALS-related genes. These advances have helped us understand the polymorphous nature of the disease, and the ambiguity between sporadic and familial forms of ALS.

Still, no known cure exists for ALS. In the last decade alone, 18 clinical and preclinical trials have failed, most recently with a drug called dexpramipexole, which cost the US-based company Biogen $100 million to develop and test. Today, only two drugs are approved by the US Food and Drug Administration (FDA) for treating ALS symptoms. The first, approved in 1995, is Rilutek (riluzole), a costly neuroprotective medication first discovered in the 1940s. It minimally improves bulbar muscle (involved in speech and swallowing) and limb function, but not general muscle strength. Rilutek is relatively safe and probably prolongs median survival by about two to three months.[4] Chinese clinicians recently confirmed its modest beneficial effect.[5]

The second drug, known as either edaravone (Radicava) or MCI-186, was first approved in Japan in 2015[6] and then approved by the FDA on May 5, 2017. An anti-oxidant that works in the central nervous system as a potent scavenger of oxygen radicals,[7] patients in treatment experienced a 33 percent reduction in the rate of decline in physical function compared with patients on placebo. In Belgium, the country where I work, a patient association called Belgian ALS League found that the intravenous injection necessary to administer the drug presents a "high burden" and renders the drug "less attractive."

A Cultural Phenomenon

The Ice Bucket Challenge was a campaign to raise money for charity that took the internet by storm in July 2014, and many supporters trace the tipping point and focus on ALS to a video posted by Pete Frates, a former college baseball player who was diagnosed with the disease in 2012. The campaign left scores of notable persons—from Bill Gates and Mark Zuckerberg to George W. Bush and Anna Wintour—shivering and drenched. More importantly, it paid off in the most spectacular way, raising $220 million worldwide for research, education, and patient support for ALS, and more than $115 million of the total for the ALS Association.

Proceeds from the campaign helped the association fund research that identified a new gene, NEK1, that contributes to the disease, the subject of a paper published in *Nature Genetics*. A year after the challenge went viral, scientists said that the money had a big effect on the gene's discovery. Says lead researcher Philip Wong, a professor of pathology at Johns Hopkins: "Without it, we wouldn't have been able to come out with the studies as quickly as we did."

The Cause of ALS

The molecular era of discovery in ALS began in 1993 with the identification of dominant mutations in a gene that produces a 153-amino acid enzyme, superoxide dismutase (SOD1), that ordinarily protects cells against damaging byproducts of metabolic processes. Specifically, this enzyme[8] catalyzes the conversion of the highly reactive and damaging chemical superoxide to hydrogen peroxide or oxygen. More than 170 ALS-causing mutations have now been identified and lie in almost every region of the 153-amino-acid SOD1 polypeptide. Mutations in the SOD1 gene are found both in sporadic and familial ALS cases.[9]

The consensus is that the disease arises not from the mutant enzymes' loss of its protective function, but rather their acquisition of toxic properties. But since the discovery of mutations in SOD1, no consensus on the nature of such toxicity has emerged. A prominent finding is that a proportion of each ALS-causing SOD1 mutant fails to fold properly, suggesting that accumulation of misfolded SOD1 may contribute to toxicity in ALS. Misfolded SOD1 forms lead to the inclusion in cells' cytoplasm of various possibly damaging substances that occur early in ALS and escalate as the disease progresses.[10]

Human Genome Sequencing

Advances in the genetics of ALS began with the sequencing of the human genome. The 2001 draft sequence by the Human Genome Project (HGP) took some 200 scientists more than a decade and cost almost $3 billion to complete.[11] Three years later, the HGP completed mapping the genome, including 99 percent of the active/translatable DNA part sequence[12] that comprises its most common form within the cell nucleus. Since then, assembly and refinement of the reference genome has allowed scientists to identify all genes—ca. 25,000—that carry the code for producing all human proteins. The research has also provided a snapshot of genetic variation, most commonly in the form of single nucleotide polymorphisms (SNPs).

The HGP demonstrates the tremendous potential value of coordinated studies to create community resources to propel biomedical research. During the past decade, the price of genome sequencing has dropped as the process has become automated and methods have improved. Now, complete genomes can be sequenced quickly and affordably, enabling the identification of genetic variants that affect heritable phenotypes, including important disease-causing mutations.

"Gene studies [relating to ALS] did not reveal much of use until about six years ago, but they are now rapidly advancing our knowledge," says Ammar Al-Chalabi, professor of neurology at King's College in London. "When I started in ALS research in 1994, the only known [genetic] cause accounted for just two percent of cases. Now, we can explain about 20 percent of cases. Through studies of individual lifestyles and the way our DNA changes during our lives, it should become easier to design effective treatments."[13]

Excellent recent studies have described the application of genetic methods in discovering new ALS-related genes.[14-16] The list (last updated in 2015), with chromosomal location of the ALS related genes, can be seen at http://alsod.iop.kcl.ac.uk/Chromosomes/chromoALL.aspx.

ALS-related Genes

As human beings we are unique; shaped by our environment and life experiences but also by our genetic make-up. Human genetic diversity is associated with phenotypic variation, notably in our physical characteristics but also in determining our susceptibility and response to disease, often as part of a multi-factorial disease process. The most common form of genetic diversity are single nucleotide differences, a spelling change in the DNA code where one nucleotide is replaced by another but other larger structural variants are also being identified.

The challenge amidst so much genetic variation—there are over 10 million single nucleotide differences alone among humans—is to define specific variants responsible for common multi-factorial diseases such as infections, diabetes, or heart disease. There have been major advances in how we define the genetic determinants of common disease over the last few years based on using many hundreds of thousands of genetic markers across the genome to look for association with disease in "genome-wide association studies." Nevertheless, a major challenge remains once a region of the genome has been implicated in disease to then define the specific functionally important causative variants.

High-throughput DNA sequencing technologies have generated large amounts of sequence data very rapidly and at a substantially lower cost. While these technologies provide unprecedented opportunities to discover new genes and variants underlying human disease, it should be stressed that these discoveries must be rigorously performed and replicated to prevent the proliferation of false-positive findings. A set of guidelines addresses the types of issues to consider in rare variant analysis.[17] These guidelines help provide objective, systematic, and quantitative evaluation of the evidence for pathogenicity. Moreover, sharing of these evaluations and data among research and clinical laboratories will maximize the chances that disease-causing genetic variants are correctly differentiated from the many rare non-pathogenic variants seen in all human genomes.

This is particularly true for ALS, in which rare and potentially pathogenic variants in known ALS genes occur in over 25 percent of apparently sporadic and 64 percent of familial patients. The recent work of Janet Cady and others showed that a significant number of subjects carried variants in more than one gene. This finding suggests an explanation for the varied age of symptom onset, and supports oligogenic inheritance (when a few genes cause or modulate a disease) as relevant to ALS pathogenesis.[18]

Genome-wide association studies, which aim to identify common variants, have been successful in characterizing novel genetic regions associated with complex traits seen in neurodegenerative diseases such as Alzheimer or Parkinson diseases.[19] A key discovery related to ALS was the genetic variation observed at the same position (locus) on chromosome 9 that causes sporadic ALS and familial ALS-frontotemporal dementia.[20] This indicates that some of the genes that may cause ALS are pleiotropic, meaning that the same genetic mutations may produce different clinical phenotypes. The determinants of this pleiotropy are largely unknown.[21]

The two sequencing methods used in ALS research are:

- Exome sequencing, also known as whole exome sequencing (WES or WXS), a technique for sequencing the expressed genes in a genome. This approach identifies genetic variants that alter protein sequences. It costs much less than whole-genome sequencing and is thus a good choice when whole-genome sequencing is not practical or necessary. It may be extended to target functional nonprotein coding elements (e.g., microRNA, long intergenic noncoding RNA, etc.)[22] Since these variants can be responsible for both Mendelian and common polygenic diseases, such as Alzheimer's disease, whole exome sequencing has been applied both in academic research and as a clinical diagnostic. It has rapidly become the primary method for the discovery of causative genes in rare diseases.

- With exome sequencing, Elizabeth T. Cirulli and others identified more than 70 distinct pathogenic mutations across several genes; their discovery may guide future efforts to functionally characterize the role of these ALS genes.[23]

- Whole-genome sequencing (WGS) delivers a comprehensive view of the entire genome. It is ideal for discovery applications, such as identifying causative variants. Whole-genome sequencing can detect single nucleotide variants, insertions/deletions, copy number changes, and identify large structural variants. WGS has identified common genetic causes of ALS and frontotemporal dementia (FTD) and transformed our view of these disorders, which share unexpectedly similar signatures, including dysregulation of common molecular players. One player so identified is mutate TBK1, which results in impaired autophagy and contributes to the accumulation of protein aggregates and ALS pathology.[24]

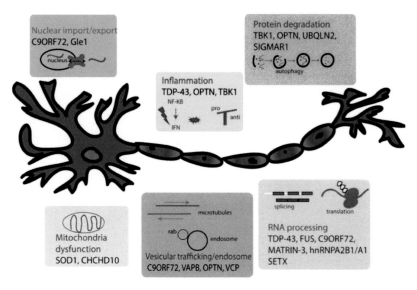

Figures 1. The following diagram illustrates some of the ALS-related genes, clustered per their role in the disease.[26] (Reproduced with the permission of Guy Rouleau, M.D., Ph.D.)[25]

Oligonucleotides and ALS Therapy

Antisense oligonucleotides (ASOs) are synthetic single stranded strings of nucleic acids, between 8 and 50 nucleotides in length, that target specific species of mRNA. ASOs inhibit gene expression or modify mutant proteins to reduce their toxicity.[26] Antisense-mediated gene inhibition was first introduced by Mary Stephenson and Paul Zamecnik in a 1978 article that has been cited more than 700 times.[27]

For dominantly inherited disorders where data suggest a mutation that confers a toxic activity on a protein (a toxic gain of function), lowering levels of the protein is a potential approach, and antisense oligonucletoides (ASOs) is one means of doing so.

In 2006, Richard Smith and others[28] pioneered the idea of ASO infusion into the nervous system as therapy for human neurodegenerative diseases. Seven years later, Timothy Miller and others,[29] using the experiments made on transgenic SOD1 mutant mice, succeeded in launching the first clinical study of delivery into spinal fluid of an ASO. The ASO was well tolerated, demonstrating the feasibility of this approach.

The most common inherited cause of ALS and FTD is the expansion of six toxic nucleotides within a gene called C9orf72. Per Genetics Home Reference, the C9orf72 gene contains a segment of DNA made up of a series of six DNA building blocks (nucleotides), four guanines followed by two cytosines (written as GGGGCC). This segment (known as a hexanucleotide repeat) can occur once or be repeated multiple times; estimates suggest repeats of up to 30 times have no negative effect on gene function. Mutations in the C9orf72 gene affect the GGGGCC segment of the gene. When this series of nucleotides is repeated too many times, it can cause ALS and/or FTD. This type of mutation is called a hexanucleotide repeat expansion. The application of ASO-mediated therapy reduced the toxicity of this hexanucleotide, further validating its feasibility.[30]

The Next Steps

From the remarkable progress already made, there is little doubt that our understanding of the genetic basis of ALS will continue to improve through conventional genetics and enhanced association studies that identify the role of rare genetic variants, including those found in non-coding DNA. However, the success of future initiatives involving the assembly of sequence and clinical data from very large-scale cohorts will require the full translation of human genetic findings into understandable biological and clinical terms. Such efforts will undoubtedly need the cooperation of multiple stakeholders, including clinicians, biochemists, informaticians, statisticians, patient participants or partners,[31] and official agencies such as the FDA.

The FDA has, in fact, taken the initiative to prepare a guidance document, "Use of Public Human Genetic Variant Databases to Support Clinical Validity for Next Generation Sequencing (NGS)-Based In Vitro Diagnostics," that describes the agency's current thinking on this subject.[32] It outlines the FDA's criteria in determining whether a genetic variant database is a valid source of scientific evidence that supports the clinical validity of an NGS-based test. Upon finalization approval of this document, test developers will be able to follow these recommendations when preparing a premarket submission—just one more step forward in the fight against a disease that must be stopped.

Acknowledgment

I would like to dedicate this article to Mrs. Marie-Henriette Burelle, surnamed Myette, who supported my research and died March 20, 2017. I thank Jacques Haiech for fruitful advice, the European Brain Council and the Belgian Brain Council executive committee for their support, and Alison Turner for having revised the English.

6

The Sleeping Brain

By Chiara Cirelli, M.D., Ph.D., and Giulio Tononi, M.D., Ph.D.

Chiara Cirelli, M.D., Ph.D., is a professor in the Department of Psychiatry at the University of Wisconsin–Madison. She received her medical degree and Ph.D. in neuroscience from the University of Pisa, Italy, where she began her investigation of the molecular correlates of sleep and wakefulness and the role of the noradrenergic system in sleep regulation. She continued this work at the Neuroscience Institute in San Diego as a fellow in experimental neuroscience, and subsequently at the University of Wisconsin. Cirelli's research is aimed at investigating the fundamental mechanisms of sleep regulation by using a combination of molecular and genetic approaches.

Giulio Tononi, M.D., Ph.D., is professor of psychiatry, distinguished professor in consciousness science, the David P. White Chair in Sleep Medicine at the University of Wisconsin–Madison, and the director of the Wisconsin Institute for Sleep and Consciousness. He received his medical degree from the University of Pisa, Italy, where he specialized in psychiatry. After serving as a medical officer in the Army, he obtained a Ph.D. in neuroscience as a fellow of the Scuola Superiore for his work on sleep regulation. From 1990 to 2000, he was a member of The Neurosciences Institute and, in 2005, received the National Institutes of Health Director's Pioneer Award for his work on sleep. His laboratory studies consciousness and its disorders as well as the mechanisms and functions of sleep.

Editor's Note: The role of sleep has long baffled scientists, but the latest research is providing new indicators about what it does for both the brain and body. While scientists believe that sleep re-energizes the body's cells, clears waste from the brain, and supports learning and memory, much still needs to be learned about the part it plays in regulating mood, appetite, and libido.

WHY DO WE SPEND A THIRD of our life asleep? The answer seems obvious: To recover from the "fatigue" of being awake, to be ready for another day of challenges, good or bad. All of us have experienced the consequences of a sleepless night: everything requires more effort; we lack energy and motivation, and feel groggy, irritable, and snappish.

But there is strong, objective proof that, far from being just such a "time filler," sleep serves an active, essential function.[1] We know that all animal species that have been carefully studied sleep, with no exception.[1] If sleep were not essential, one would expect that some would have evolved to do without it, since time spent asleep reduces time spent foraging, reproducing, or monitoring the environment. Moreover, being asleep puts an animal in a potentially dangerous situation, because it reduces the ability to promptly respond to stimuli that signal threat. Thus, sleep makes little sense, from an evolutionary point of view, unless it provides enough essential benefits to overcome its inherent risks.

We also have learned that sleep is exquisitely regulated. There are complex mechanisms in our brain that increase the duration and/or the depth of sleep after sleep deprivation[2] to permit the recovery of some of what was lost (homeostatic regulation). There are also neural mechanisms that tend to consolidate sleep at a certain phase of the 24-hour cycle (circadian regulation): night for us and day for many rodents. Sleep loss or chronic sleep disruption also has many negative consequences, including adverse effects on metabolism and immune function.[3,4]

The most obvious of these adverse effects are on the brain. Cognitive deficits of many kinds[5] are apparent in humans, although with substantial inter-individual differences[6,7] after just one night of total sleep deprivation or when sleep is cut short by several hours every night for a week or more.[8,9] Attention, working memory, and the ability to learn and remember decline. When we are sleep deprived, it is more difficult to speak fluently, assess risks, and appreciate humor.[8,9] Importantly, experiments have shown that these cognitive impairments can be reversed by sleep but not by the same period of quiet wakefulness.[10] Similarly, there is evidence that cognitive deficits caused by sleep loss at night can be prevented or delayed by naps.[11] The open question is: Why does sleep offer a special time for the brain's recovery?

Sleep versus Rest

In our search for an answer, we started by identifying the fundamental feature of sleep—what distinguishes it even from quiet, restful wakefulness: sensory disconnection. During quiet wakefulness, when we sit on a sofa in a silent and dark room after having exercised, for example, our muscles can recover from fatigue. Yet, we are still able to react and move promptly if the phone rings. In other words, we are still connected to the world. On the other hand, when deeply asleep, our capacity to react to a mild stimulus—a noise coming from the next room, or that phone call—is reduced substantially.

Thus, any hypothesis about the essential function of sleep must take into account that, when asleep, we are essentially offline: sensory disconnection must be an essential requirement for whatever function sleep serves. If not, natural selection would likely have found a way to perform the same function while awake, avoiding the danger of being unable to monitor the environment.

Over the past 20 years, we have developed and tested a comprehensive hypothesis about the core function of sleep: The Synaptic Homeostasis Hypothesis (SHY).[12-14] Briefly, SHY states that sleep is the price we pay for brain plasticity. According to our hypothesis, during wakefulness, synapses—the links that allow neurons to communicate with each other—undergo net strengthening (potentiation) as a result of learning. Learning is an ongoing process that happens all the time while we are awake, constantly adapting to an ever-changing environment. The remarkable and pervasive plasticity of the brain is an established fact and is essential for survival. However, it is a costly process, because stronger synapses increase the demand for energy and cellular supplies, lead to decreases in signal-to-noise ratios (because neurons would start responding less selectively to stimuli), and saturate the ability to learn.

This is where sleep enters the picture. According to SHY, while our brain is offline—disconnected from the environment—neural circuits can be reactivated, renormalizing synaptic strength. As explained below, this renormalization favors memory consolidation and the integration of new with old memories, and eliminates the synapses that contribute more to the "noise" than to the "signal." Just as importantly, synaptic renormalization during sleep restores the homeostasis of energy and cellular supplies, including many proteins and lipids that are part of the synapses, with beneficial effects at both the systems and cellular level.

Thanks for the Memories

It is useful to expand a bit on the rationale for SHY. First, the hypothesis emphasizes the obvious fact that wakefulness is the time when we can learn about our current environment, since only then do correlations in a neuron's input reflect the causal structure of that environment. By the same token, sleep is not a good time for forming new memories, since we would run the risk of remembering fantasies and dreams, rather than events in the real world.

Second, learning is massive and ubiquitous throughout wakefulness: every waking minute you lay down innumerable neural traces about facts and events. Try the simple experiment of recalling everything that you saw, heard, and did today. You will realize that it is possible to remember innumerable facts and events about your daily activities, even if you did not start the day with the specific goal of memorizing them.

Third, learning during waking should occur primarily by synaptic potentiation, not depression; i.e. by strengthening rather than weakening the connections among neurons. Why? Because firing is much more energetically expensive than silence[15,16] by default neurons that tend to fire very little.[17,18] And since any decrease in firing starting from such low levels would be hard to detect, neurons should signal an important event by increasing their firing.

So how can a neuron, located deep in the brain without direct access to the external world, judge what events are potentially important? Simply by paying attention to those inputs (or synapses), out of the few thousands it receives, that are firing strongly and together. The increased firing of several inputs, more or less synchronously, represents a "suspicious coincidence" and is a good indicator that many of its input neurons had something important to signal, hence worth relaying further. During a waking day, then, one can expect that many synapses will be strengthened throughout the brain, establishing memory traces.

But now a different issue arises. If the brain undergoes massive synaptic potentiation to establish memory traces during a waking day, there must be a way to ensure that synaptic strength does not grow indefinitely. Otherwise, neurons would quickly reach synaptic saturation, obliterating all memory traces, not to mention that energy and supplies would not be sustainable under these conditions. Clearly, there must be synaptic renormalization to regulate synaptic strength; a kind of synaptic homeostasis to prevent the nervous system from descending into chaos.

In this respect, the default assumption among neuroscientists has been that synaptic homeostasis is maintained during learning itself, when the brain is online. SHY, instead, proposes that synaptic renormalization should not happen during waking, when we are at the mercy of a particular environment and slaves of the "here and now," but during sleep, when the brain is offline. Freed from the tyranny of its immediate environment, the brain can sample all its memories—old and new—and renormalize the total amount of synaptic strength in a smart way, preserving and consolidating those newly formed memory traces that fit best with its overall knowledge basis, while forgetting those that fit less well. Thus, sleep should be a time for net synaptic depression, leading to optimal "down-selection" of memory traces.

Testing a Hypotheses

How can the general idea underlying SHY be tested? There is no one comprehensive measure of synaptic strength, especially in vivo in freely moving animals. Over the years, we have employed many different methodologies, from continuous electrophysiological recordings in rodents and humans to molecular and genetic experiments in flies, mice, and rats. For instance, by using electrical stimulation (in rats) or transcranial magnetic stimulation (in humans, using a magnet above the skull to stimulate the brain non-invasively), we have found that the neural response triggered by a stimulus of fixed amplitude is bigger after several hours spent awake compared to several hours spent asleep,[19,20] consistent with a net increase in synaptic strength during wakefulness.

These results are in line with SHY's predictions, but one cannot rule out that changes in neuronal excitability may also play a role.[21] More direct evidence comes from experiments in cortical slices, where we found that frequency and amplitude of miniature excitatory synaptic potentials increase after waking and decrease after sleep.[22] Measuring the total amount of synaptic proteins in large regions of the fly and rat brain, we similarly found an increase after waking as compared to after sleep.[19,23,24] Furthermore, in adolescent mice in which neural circuits are still being formed and refined, we found that during wakefulness the formation of new synapses exceeds their elimination, while the opposite is true during sleep.[25]

Perhaps the most direct way to measure synaptic strength is through an ultrastructural approach that literally measures the size of the synapses. It is well established that stronger synapses are bigger, which implies that, per SHY, most synapses should grow with wakefulness and shrink with sleep, in a matter of a few hours. To test this prediction, we used serial block face electron microscopy (SBEM), a new method that permits the effective and automatic acquisition of high-resolution, tridimensional images of many synapses. Because the actual measurements of the synapses must be performed manually, it took several people four years to measure ~7,000 synapses in the mouse cerebral cortex.

At the end of this laborious process, the results were clear: six to eight hours of sleep led, on average, to an 18 percent decrease in the size of the synapses as compared to six to eight hours of non-forced wakefulness at night. Similar results were obtained when sleep was compared to six to eight hours of wakefulness enforced by exposure to novel objects during the day,[26] proving that these changes are due to behavioral state (sleep vs. waking) and not to circadian factors (day vs. night). The sleep-related decline in synapse size occurred in both areas that were examined—primary motor and primary sensory cortex.

Across the population of neurons, the decrease in synaptic size with sleep largely followed a scaling relationship, meaning that the change was proportional to the size of the synapse. Intriguingly, however, sleep-related downscaling showed some selectivity, occurring in the great majority of synapses (~80 percent) but sparing the largest ones, which may be associated with the most stable memory traces. The magnitude of these changes is worth considering. There are approximately 100 billion neurons in our brain, 16 billion of them in the cerebral cortex alone (14 million in the mouse cortex), and each neuron receives thousands of synapses. Thus, if what we observed in two cortical areas extends to other brain regions, every night trillions of synapses in our brain could get slimmer by nearly 20 percent.

The SBEM study was recently published along with research from an independent group, which used biochemical and molecular methods to confirm SHY's prediction that synapses undergo a process of scaling down during sleep.[27] This study also showed that one gene, Homer1a, is important for the sleep-mediated downscaling process.

Avenues to Explore

We still do not know how synaptic renormalization occurs at the individual synapse level, which would require measuring changes across the sleep/wake cycle and before and after learning. SHY proposes that total synaptic strength is downscaled during sleep, but this does not mean that all synapses need to be renormalized every night. In computer simulations, we showed that various synaptic rules enforcing activity-dependent depression during

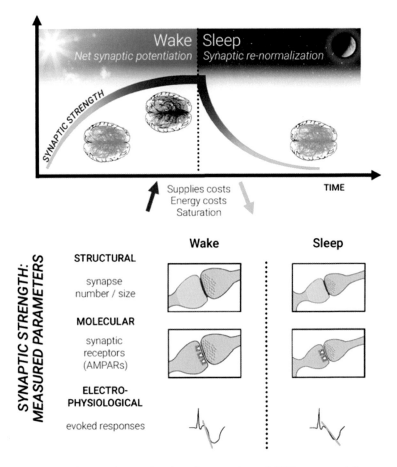

Figure 1. Top, Schematic diagram describing the main claim of SHY: net synaptic increase occurs during wake (during the day in humans and diurnal animals), when many circuits in the brain get potentiated (dark blue lines in the brain schematic), resulting in cellular and systems' costs, followed by synaptic renormalization during sleep, when most, if not all circuits undergo synaptic down-selection (light blue lines).

Bottom, Parameters used to test SHY. While structural and molecular measures more directly reflect synaptic strength, electrophysiological measures such as responses evoked by direct stimulation of the cerebral cortex can be strongly affected by other factors that modulate intrinsic neuronal excitability, including the levels of neuromodulators and the balance between excitation and inhibition, and thus cannot be used alone to infer synaptic strength. In the upper panel (structural), the red line indicates the axon to spine interface (ASI), and the yellow area outlines the head of the spine. In the middle panel (molecular), excitatory glutamatergic receptors (AMPARs) are shown as squared boxes.

sleep are compatible with the renormalization process predicted by SHY.

For example, it could be that stronger synapses are either depressed less than weaker ones[28] or are completely protected from depression.[29,30] In the latter implementation of down-selection, when a neuron detects increased firing in many of its inputs during sleep (and thus fires strongly), the associated synapses will be protected from depression and maintain their current strength. The result is that, while the strength of these synapses does not increase in absolute terms, as is the case during wake, it does increase in relative terms, because the remaining synapses, those that were not protected, end up losing strength. This competitive down-selection mechanism has the advantage that synapses activated strongly and consistently during sleep survive mostly unchanged and may actually consolidate, becoming more resistant to interference and decay.

We believe that through this single process—synaptic renormalization—sleep can provide many of the benefits that behavioral studies have documented at the systems level: The ability to learn new things the next day, the consolidation of what has already been learned, and "smart" forgetting (i.e., the systematic elimination of irrelevant memory traces). However, much remains to be done. We plan to test whether the ultrastructural changes found in the cerebral cortex also happen elsewhere in the brain, such as in the hippocampus, another region crucial for brain plasticity and learning.

It is also essential to establish what happens during early development, when brain circuits are still being formed and synapses show a very high turnover. Additionally, we are trying to determine what might happen during sustained sleep loss: Do synapses keep growing as in normal waking, or do they start shrinking and disappearing, perhaps through a massive "pruning" process that is not as carefully regulated as during normal sleep? And if so, is abnormal pruning more pronounced during sensitive periods of development when neural circuits are still forming, perhaps with long-term consequences for the wiring of the adult brain?[31,32]

7

The Brain's Emotional Development

By Nim Tottenham, Ph.D.

Nim Tottenham, Ph.D. is an associate professor of psychology at Columbia University and director of the Developmental Affective Neuroscience Laboratory. Her research examines brain development underlying emotional behavior in humans, and highlights fundamental changes in brain circuitry across development and the powerful role that early experiences, such as caregiving and stress, have on the construction of these circuits. Tottenham has authored over 80 journal articles and book chapters and is a frequent lecturer both nationally and internationally on human brain and emotional development. She is a recipient of the National Institute of Mental Health Biobehavioral Research Award for Innovative New Scientists (BRAINS), the American Psychological Association's Distinguished Scientific Award for Early Career Contribution to Psychology, and the Developmental Science Early Career Researcher Prize.

Editor's Note: From our earliest days, the brain rapidly develops thinking, mobility, and communication skills. But not quite as quick to develop are the parts of the brain that regulate and process our emotions. New research is helping scientists learn about areas that are crucial to emotional development, and how our surroundings fit into the picture. The findings could have far-reaching implications for both parents and policy-makers.

HUMANS ARE LIKELY THE most emotionally regulated creatures on earth. Compared to other animal species, we can modulate and modify emotional reactions and experiences, even very intense ones, through a large and sophisticated emotion regulation repertoire that includes skills of distraction, reappraisal, language, prediction, social interaction, suppression, and more.[1-5] At times, these skills require effort, and at other times, they seem reflexive and automatic.

But what are some of the variables in this sophisticated emotion regulation repertoire? The parent of any toddler or even adolescent can attest to the very slow development of emotion regulation processes. This slow development has been documented in empirical research, which also notes the large individual differences from one person's ability or style of emotion regulation to another's.

Evolutionarily speaking, this slow development of emotion regulation ability in childhood that culminates in an exquisite ability in adulthood points to the benefits of a slow-maturing emotion regulation system. Indeed, humans are not only a highly emotionally regulated species, but they are slowly developing in general, relative to other species,[6] with a prolonged period of immaturity. Phylogenetically, slow development may confer benefits through an extended period of neural plasticity—a feature of a developing neural system that heightens its ability to learn from the environment. If so, then humans may owe their sophisticated emotion regulation skills to the "extension" of childhood that has evolved in us.

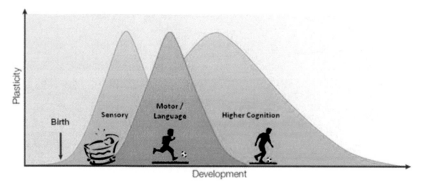

Figure 1. Windows of plasticity in brain development

The nature, chronicity, and quality of environmental inputs during these periods of plasticity, in particular those from close relationships (e.g., parents, friends, teachers), in large part determine emotion regulation functioning in adulthood.[7] Thus, adult brain and behavioral function in this regard can be conceptualized as a historical reflection of what was experienced during development. To fully appreciate individual differences in adult emotion regulation skills, then, it is helpful to understand how the brain develops.

Emotion Regulation in Adulthood

While it is probably fair to say that much of the brain contributes in some form or another to emotion regulation, at its core the processes rely on communication between areas of the prefrontal cortex, in particular medial regions and subcortical systems including the amygdala, hippocampus, and basal ganglia.[8-10] The medial prefrontal cortex (mPFC) includes association cortex (meaning that it can synthesize incoming information from multiple sources) and has strong bidirectional connections, which typically occur via a single synapse,[11-14] to and from subcortical regions. The mPFC receives and coordinates signals from perceptual, semantic, and linguistic regions of the brain as well, to facilitate regulation—e.g. amplification, redirection, or dampening—of emotional reactions.[5]

Prefrontal Cortex and Sensitive Periods

This complex neurobiological interstate takes years to reach maturity. It is often the case that slow developing systems are highly susceptible to environmental pressures; that is, they exhibit a high degree of plasticity. In a given neural system's development, there is a moment when it is particularly sensitive to the environment—a so-called sensitive period. The brain undergoes multiple sensitive periods, a different one for each neural circuit, and   taken together such intervals span development. Sensitive periods, and their less-forgiving cousins, critical periods, have been studied more often in perceptual systems (e.g., vision) than in emotion regulation. Thus, we will illustrate the properties of sensitive periods with an example from vision development. Then we will discuss how these principles may apply to emotion regulation.

The Nobel prize-winning work of Hubel and Wiesel[15] demonstrated that a window of heightened brain plasticity opens in the early days of a cat's visual system; during this window, environmental input (i.e., light) leaves an enduring influence on the visual cortex and associated visual behavior (i.e., binocular vision) that is impossible to reverse once the critical period terminates. That is, once the window of heightened plasticity closes (i.e., the cat becomes older), the developing system is no longer modifiable by light.

More recently, this pioneering work has been extended in rodent models to reveal the molecular mechanisms that give rise to the opening and closing of these windows of plasticity.[16,17] This research identified core cellular and molecular mechanisms, highlighting the central role of shifting inhibitory and excitatory neurotransmitter input balances (e.g., GABA and glutamate) across development. In addition, it discovered the environmental means through which sensitive periods themselves might be altered, in terms of timing, duration, and magnitude.[7] It suggested, in other words, that the nature of the sensitive periods themselves might be plastic and modifiable by early experiences (e.g., sensory deprivation, high exposure to video games, maternal depression[7]).

Such periods of heightened plasticity (and the accompanying rapid learning) have been identified in humans. For example, using clever behavioral studies with preverbal infants, researchers have discovered that within the first six months of postnatal life, the infant's auditory system becomes attuned to the sounds (i.e. phonemes) that are meaningful in the language to which the infant is exposed.[18,19] Similar effects have been identified in human face recognition.[20]

While sensitive/critical periods have been identified for the development of perceptual systems, their existence in higher-level processes, such as emotion and cognition, has remained elusive. One possibility is that there are no sensitive periods for these processes and that sensitive periods are an exclusive property of perceptual systems. Another is that sensitive periods exist for higher order brain processes (e.g., cognition, emotion), just as they would for any developing neural system, but exhibit much broader and less well-defined temporal boundaries than earlier-developing perceptual systems.

The latter is what we might expect from very slow-maturing regions of the brain, such as the prefrontal cortex. Indeed, a large volume of literature spanning the past two decades has demonstrated that developmental change in this region continues well into adulthood, both structurally and functionally,[21-24] far outpaced by perceptual systems[25,26] and lower subcortical structures, including those involved in emotional learning.[27-32] This pattern reflects the hierarchical nature of development, the process whereby a system's functioning depends upon the prior maturation of other systems.

The slow development of emotion regulation is paralleled by the slow development of the neurobiology that supports it (e.g., the amygdala and mPFC). Prolonged age-related change in prefrontal cortex and subcortical connectivity has been demonstrated both by structural and functional connectivity measures.[27,31-38] Several studies have shown that communication between mPFC and the amygdala, which supports emotional learning and arousal, is qualitatively different in childhood than in adulthood (even though continued quantitative change is observed throughout adolescence and young adulthood). Specifically, in adulthood (but not earlier), increases in mPFC activity are associated with a decrease in the amygdala's activity— these two regions are anti-correlated with each other in response to emotional stimuli (such as fear faces).

This anti-correlation, which has been supported by more invasive studies involving lesions in the brain,[39] is interpreted to reflect mPFC regulation of the amygdala in healthy adulthood—i.e., top-down information flow from mPFC to amygdala. Unsurprisingly, adults who typically exhibit this pattern of anti-correlated amygdala and mPFC communication in response to emotional cues are those with better emotion regulation. On average, children do not show this adult-like pattern, but instead, a child-specific one. [27,37,40,41]

This child-specific pattern is a compelling argument for a sensitive period in the development of connections between the amygdala and mPFC. One should note that It can be quite challenging to definitively identify sensitive periods in humans, because our period of development takes so long, posing considerable difficulties in experimental control and design. This is where translation from animal studies becomes necessary. For example, the period between weaning and puberty has emerged as the sensitive period for mPFC-amygdala connections in the rodent.[42] Precise tracing and optogenetic studies have shown that amygdala-to-mPFC connections develop earlier than regulatory connections in the opposite direction.[43-46] Consequently, the neurobiology of emotion regulation differs qualitatively between late infancy and adolescence—where the adolescent rat uses the mPFC for regulating emotion, the infant does not.[47]

Similar findings have been observed in human fMRI studies.[48] The connections between mPFC and amygdala exhibit massive development during the post-weaning/pre-pubertal period, and the boundaries of this change are marked by drastic shifts in the excitatory/inhibitory balance—the molecular hallmark of sensitive period opening/closure.[43] Moreover, stimuli (e.g., music) learned during this juvenile period have enduring consequences on mPFC-amygdala function observed throughout adulthood, and this learning must occur during a sensitive period.[42] In 2016, Laurel Gabard-Durnam and others showed that in humans, the nature of the co-activation of mPFC and amygdala in response to environmental stimuli predicts the development of future functioning most strongly when it is measured in childhood.

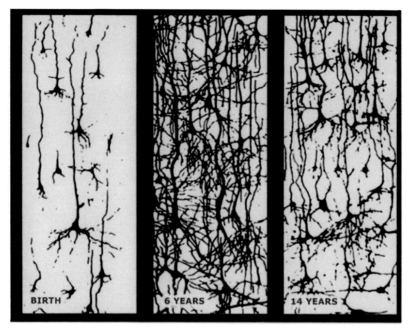

Figure 2. The postnatal development of the human cerebral cortex; more than one million new neural connections per second. Image source: Colel, JL, Cambridge, Mass: Harvard University Press

Emotion Regulation without a Mature Prefrontal Cortex

Taken together, non-human and human research has established that the young animal (rodent or human) does not use the mPFC for emotion regulation in the way that the mature one does. Nonetheless, young animals can exhibit regulated emotional behavior. So how is the young animal accomplishing this task?

While the infant rat displays some forms of emotion regulation (for example, extinguishing a learned fear), the nature of the behavior is starkly different from the adult.[47] For an animal dependent on an adult for survival, there is a species-expectation that the attachment figure will be available. This caregiver can serve as an external social regulator at a time of mPFC immaturity. Indeed, the behavioral literature has noted across decades of empirical work that the young child relies on the parent in just this way, using various strategies.

A common example is social referencing.[49-51] Children routinely look to the parent for guidance in navigating the emotional and physical landscape. Social referencing is a powerful means of regulating emotions and has been used to explain the intergenerational transmission of emotional knowledge, including the transmission of anxious behaviors and reactions.[52,53]

Another mechanism by which parents can regulate emotions is through modulation of stress reactivity and fear learning. Evidence for parental buffering has been identified across various species, including rats, guinea pigs, monkeys, and now humans.[54-56] Researchers have found that there are sensitive periods in development when access to parental cues provides a powerful external means of emotion regulation. As has been described extensively in other sources,[54,57,58] stimuli related to a (regulated) parent can modulate a physiological response to threat, namely that parental cues can dampen elevations in the stress hormone cortisol (in humans) and corticosterone (in rats).[59-61]

Parental cues have also been shown to decrease amygdala activity in both humans[55] and in rats.[60] For example, in the developing rodent, this reduction of amygdala activity by the parent prevents threat learning. That is, the presence of parental cues prevents the developing rat from learning to associate a threatening stimulus with a cue (e.g., a tone or a light). Conversely, when a parent expresses defensive behaviors or heightened negative affect, the rodent work has shown that such cues are quite effective in amplifying amygdala reactivity.[62] Emerging research in humans has also demonstrated that parental cues are effective in reducing amygdala activity.[55] Taken together, these studies show that during a time of relative mPFC immaturity, the parent can serve as an external regulator of subcortical arousal. Moreover, the effect of parental cues not only provides instruction to prefrontal cortex development, but may also contribute to the high degree of plasticity observed in this region.[63,64]

We now know that the prefrontal cortex is one of the last brain regions to develop, and its connections with other cortical and subcortical targets are very slow to form. These processes are especially slow in the human, and evidence of continued development has been documented through adolescence and adulthood. This slow-paced and sustained development renders the prefrontal cortex and its connections vulnerable to environmental insults (e.g., early psychosocial adversity), but at the same time offers great potential for extensive learning from positive, enriching environments, and the optimization of neural processes that will facilitate regulated behavior. Its end-product is an incredibly rich emotional regulation repertoire in the mature adult.

8

Olfaction: Smell of Change in the Air

By Richard L. Doty, Ph.D.

Richard L. Doty, Ph.D. is the director of the University of Pennsylvania's Smell and Taste Center and the inventor of the University of Pennsylvania Smell Identification Test (UPSIT), a standardized olfactory test heralded as the olfactory equivalent to the eye chart. Dr. Doty is editor of the *Handbook of Olfaction and Gustation* (Marcel Dekker, 1995, 2003, 2015), the largest collection of basic, clinical, and applied knowledge on the chemical senses ever compiled in one volume. He is an author or co-author of over 400 professional publications (including 10 books and contributions to such publications as *Science*, *Nature*, and the *Encyclopedia Britannica*). His most recent books are *The Neurology of Olfaction*, co-authored by Christopher Hawkes (Cambridge University Press) and *The Great Pheromone Myth* (Johns Hopkins University Press). Among his numerous awards are the James A. Shannon Award from the National Institutes of Health (1996), the Olfactory Research Fund's Scientific Sense of Smell Award (2000), the William Osler Patient-Oriented Research Award from the University of Pennsylvania (2003), and the Association for Chemoreception Science's Max Mozell Award for Outstanding Achievement in the Chemical Senses (2005).

Editor's Note: Every whiff you take not only brings a cloud of chemicals swirling up your nose, but matters to your experience of taste as well as smell. Scientists studying smell have not only provided compelling evidence that it's more sophisticated than previously thought, but believe that the sense of smell impacts our mood and behavior and has the potential to detect and treat some neurological disorders. Compared to other senses, smell has long been underappreciated, writes our author, but that is now beginning to change.

IN MANY WAYS, THE SENSE OF SMELL—also known as olfaction (from the Latin word for smell or odor, *olfactorius*)—is our most complex sensory system, capable of distinguishing thousands, if not millions, of different odors, often at concentrations below those detectable by sophisticated instruments. The biological machinery required for such a feat is remarkable, involving six to ten million odorant receptor cells embedded within a thin layer of tissue (epithelium) high in the nasal chambers. Each receptor cell projects 10 to 30 thread-like cilia into the mucus covering the epithelium (see figure 1 on following page).

These cilia carry receptors—protein structures specialized to respond to odor-carrying molecules, termed odorants. There are nearly 400 types of receptors, different subsets of which are triggered by different odorants. Each of the olfactory receptor cells contains only one type of receptor. When enough odorant molecules get through the mucus layer and bind to the receptors located on the receptor cell cilia, they stimulate a neural impulse. The pattern of activated receptors is specific to a given odorant, and the neural impulses that are generated are decoded by higher brain regions. Memory processes also become involved, allowing us to recognize odor sensations of which we are familiar.

The major breakthrough for understanding the first phase of olfactory transduction came in 1991 when Linda Buck and Richard Axel at Columbia University published their seminal paper identifying the gene family responsible for the expression of the olfactory receptors.[1] Their work, which led to the 2004 Nobel Prize in Medicine or Physiology, put to rest questions as to the nature of initial olfactory transduction that had puzzled scientists for hundreds of years.

Since then the situation has become much more complex, with ever more types of receptors being found on some olfactory receptor cells,[2] and the realization that olfactory receptors are present in tissues throughout the body. These include the skin,[3] skeletal muscle,[4] thyroid,[5] heart,[6] lungs,[7] pancreas,[8] thymus,[9] prostate,[10] kidney,[11] bladder,[9] testes,[5] intestine,[12] blood vessels,[13] and ganglia of the autonomic nervous system.[13] Their widespread distribution raises the question of whether these receptors should, indeed, be called "olfactory." Depending upon their location, they are involved in many biological and physiological processes, including glucose homeostasis, lung ventilation, regulation of blood pressure, mitigation of tumor progression, promotion of angiogenesis, facilitation of sperm motility, induction of wound healing, and alteration of gut motility and secretion.[14] Understanding the role of such receptors in these diverse processes is opening new vistas for novel pharmacological approaches to disease management and treatment.

Figure 1. A transition region between the human olfactory neuroepithelium (bottom) and the respiratory epithelium (top). Arrows identify olfactory receptor cell dendritic endings with cilia, some of which are cut off. (From Menco and Morrison. 49 Copyright © 2003 Richard L Doty.)

The Underrated Sense

Although most of us take our sense of smell for granted, it is critical for our well-being. We use olfaction to verify the cleanliness of our clothes and homes, and to fully enjoy foods, beverages, personal care products, flowers, and other aspects of our environment. Without a sense of smell, we are exposed to the dangers of spoiled food, leaking natural gas, burning electrical wires, smoke, and other environmental hazards. The sense of smell is so important that those who can't smell (anosmics) are disqualified for "appointment, enlistment, and induction" into the US Armed Forces. Indeed, the lack of smell ability can be the basis for retirement or discharge.[15]

Multiple factors influence our ability to smell. These include occupation (e.g., perfumers and master sommeliers perform better than most of us on olfactory tests), sex, age, exercise, smoking, nutrition, head trauma, disease, and exposures to viruses, bacteria, and xenobiotics.[16] The influence of age on smell function is particularly salient. More than three-quarters of

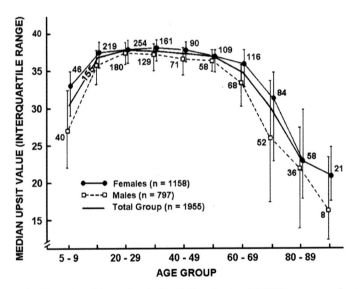

Figure 2. University of Pennsylvania Smell Identification (UPSIT) scores as a function of age and sex. Note that women identify odorants better than men at all ages although significant overlap occurs. Numbers by data points indicate sample sizes. (From Doty and colleagues.[17] Copyright 1984 American Association for the Advancement of Science.)

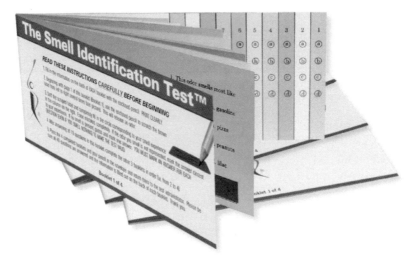

Figure 3. The University of Pennsylvania Smell Identification Test can be self-administered and is comprised of four booklets, each of which contains ten 'scratch and sniff' odorants. The subject's task is to identify the smell of each odorant based upon cued alternatives. Norms based upon 4,000 subjects are available for this test which has been translated into 30 different languages and administered to over one million people. (Photo courtesy of Sensonics, International. Copyright © 2004, Sensonics International, Haddon Heights, NJ.)

individuals over the age of 80 have a demonstrable smell problem, usually reduced sensitivity.[17] This decrease is illustrated below (see Figure 2), where a major drop occurs in later life in scores on the University of Pennsylvania Smell Identification Test (UPSIT), a 40-odor smell test developed at our center in the early 1980s (Figure 3).[18] (Note in Figure 2 that women, on average, outperform men on this test and maintain their superiority into later life.) Over the age of 85 years, 40 percent of men and 26 percent of women are anosmic, i.e., have no sense of smell at all.[19]

Taste versus Smell

It is noteworthy that many patients who come to our center's clinic for evaluation and treatment do not recognize that they have a smell problem, complaining only of a "taste" problem. However, when we formally test them, their ability to smell proves to be the culprit. Why? Because the taste buds, located throughout the oral cavity but primarily on the tongue, sense only sweet, sour, bitter, salty, and savory ("umami) sensations. Everything else we think of as taste, such as strawberry, apple, chocolate, coffee, butterscotch, meat sauce, and apple pie, are really *flavors* that depend upon the sense of smell.

This is easily demonstrated. The next time you sip your coffee, pinch your nose closed while swishing the coffee around in your mouth. You will notice the bitterness of the coffee, its warmth and smoothness, but not the distinctive coffee flavor. Indeed, the sensation is much like swishing bitter hot water. The reason why the coffee "taste" disappears is that, by holding your nose shut, you block the flow of coffee flavor molecules from the rear of the oral cavity to the olfactory receptors via the nasopharyx, the opening from the mouth into the nose.

I was once asked to visit a large retirement home to better understand why there was concern about bad tasting food. The situation was quite dire, since some residents had largely refused to eat the food and a few were wasting away. Such turmoil arose around this issue that many of the residents were on a campaign to have the chef fired. The chef eventually quit, but the problem continued even after a new chef had come on board. I had the opportunity to give a lecture to 100 or so of the residents on nutrition and the chemical senses, and passed out the UPSIT for them to take immediately after my talk. It was at this time that most of the residents discovered they could not smell very well, with most exhibiting hyposmia (decreased smell function), and that the problem with the institution's food was largely in their noses, not in the kitchen!

Smelling Danger

Aside from altered perception of the flavor of foods and attendant nutritional issues, the loss of smell function that most of us experience in our later years has potentially lethal consequences. This is evidenced by the disproportionate number of elderly people who have died in accidental gas poisonings, in part because of their inability to detect the odor added as a safety factor to natural gas.[20] Most of us have, at some time in our lives, used this sense to avert a danger from something burning on the stove, a smoldering wire in an electrical outlet, or gas from a stove that was not completely turned off. In a study of over 1,000 persons conducted by the Medical College of Virginia, anosmics were three times more likely than those with a normal sense of smell to report having experienced a potentially life-threatening event such as the ingestion of spoiled food or a failure to detect smoke or leaking natural gas.[21]

Recent studies suggest that smell loss is an extremely strong risk factor for death in healthy older persons, increasing the odds of dying by more than 300 percent over a four-to five-year period.[22] Although the reasons behind the association have yet to be determined, smell loss is a stronger predictor of death than cognitive deficits, cancer, stroke, lung disease, or hypertension, even after controlling for sex, age, race, education, socioeconomic status, smoking behavior, cardiovascular disease, diabetes, and liver damage.[22, 23]

Independent of age, we now know that smell dysfunction heralds the onset of a number of neurological diseases, including Alzheimer's and Parkinson's diseases.[24] In some cases, such dysfunction occurs years, even decades, before the appearance of the classic disease symptoms. The olfactory bulb—a structure at the base of the brain that receives information from the olfactory receptor neurons—is among the two brain regions that first show disease-related pathology in Parkinson's disease.[25] Although damage to the olfactory bulb also occurs in early Alzheimer's disease, some studies suggest that disrupted connections between the olfactory cortex and the hippocampus, which is involved in memory, predate it.[25] Factors including damage to neurotransmitter systems critical for olfactory function may precede the development of the neuropathology of such diseases, possibly even catalyzing disease development itself.[26]

Impact on Memory

We and others have observed a close relationship between the ability to smell and memory: older persons with olfactory loss are more likely to report difficulties in memory.[27] In a study of 1,092 non-demented elderly persons (average age 80 years) from a multi-ethnic community in New

York, UPSIT scores varied with the degree of memory impairment and with performance on a number of cognitive tests.[28] Importantly, such scores were weakly correlated with the MRI-determined volume of the hippo-campus, a brain structure intimately associated with memory.

Numerous longitudinal studies have demonstrated that olfactory defi-cits are associated with future cognitive decline and Alzheimer's disease in older populations.[29] Moreover, interactions with genetics have also been demonstrated. In 1999, Amy Bornstein Graves and her associates at the University of South Florida published a pioneering study in which a 12-odor version of the UPSIT was administered to 1,604 community-dwelling senior citizens. None initially showed any signs of dementia.[30] The smell test scores were found to be a better predictor of cognitive decline over the study's two-year period than scores on a global cognitive test. Persons with normal smell function who carried one or two APOE-4 alleles (a risk fac-tor for Alzheimer's disease) had only a slight elevation in risk of developing cognitive decline. However, anosmic carriers of such alleles had a five-fold risk in developing cognitive decline during this time period. While this risk was approximately three-fold in men, it was nearly ten-fold in women.

Close relatives of patients with Alzheimer's disease appear to have poor-er smell function than non-relatives of the same sex and age, implying that genetic factors may be at play.[31] However, the potential influences of inter-actions between environmental factors and genes, including the APOE-4 allele, cannot be overlooked. Using the UPSIT, Calderon-Garciduenas and her colleagues in Mexico City found olfactory dysfunction in 35.5 percent of 62 young persons (average age 21 years) living in a highly polluted area of that city, as compared to only 12 percent of those living in a much less polluted city.[32] Interestingly, APOE-4 carriers from the polluted areas failed to identify 24 percent on the 10 items reported to be particularly sensitive to Alzheimer's disease,[33] whereas carriers of the APOE 2/3 and 3/3 alleles, which are not risk factors for Alzheimer's disease, failed only 13.6 percent of such items, suggesting that APOE-4 gene carriers are more susceptible to adverse influences of pollution on their olfactory pathways.

Association with Parkinson's

In the 1990s, G. Webster Ross and his collaborators at the University of Hawaii administered the 12-item version of the UPSIT to 2,276 non-symptomatic elderly men (average age: 80) of Japanese ancestry.[34] Over the next four years, 35 were clinically diagnosed with Parkinson's disease, a disorder whose motor dysfunction is due to damage to areas of the brain that employ the neurotransmitter dopamine for motor control. After adjusting for age, smoking behavior, and other confounders, those persons whose initial olfactory test scores fell within the bottom 25 percent of the group were five times more likely to develop Parkinson's disease than those whose test scores fell within the top 25 percent.

Additional support for early olfactory involvement in Parkinson's

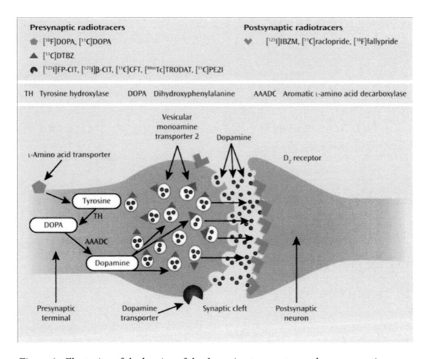

Figure 4. Illustration of the location of the dopamine transporter on the pre-synaptic membrane of a dopaminergic cell. The dopamine transporter removes dopamine from the synaptic cleft and returns it to the terminal of the presynaptic nerve cell. From Fusar-Poli and colleaguesl.[50] Copyright © 2012 American Psychiatric Association.

Control Subjects   **PD Subjects**

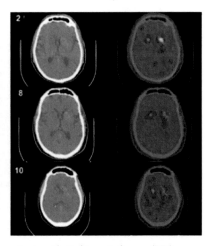

Figure 5: Imaging of the dopamine receptor in patients with Parkinson's disease (PD) (Right) and age-and sex matched normal controls (Left). Note the depletion of the radioactively labelled ligand sensitive to the dopamine transporter in the PD subjects relative to the controls. Photograph courtesy of Jacob Dubroff, Smell and Taste Center and Department of Neuroradiology, University of Pennsylvania.

disease came from a 2004 study performed at the Vrije Universiteit (VU University) in Amsterdam. In this study, olfactory tests were administered to 361 asymptomatic first-degree relatives of Parkinson's disease patients.[35] The health of regions of the brain involved in motor control also was determined using single photon emission tomography (SPECT) imaging. This procedure measures the concentration of dopamine transporter, the protein responsible for returning the neurotransmitter into cells after its release into the synaptic cleft (see Figure 4). Those with the best (top 10 percent) and worse (bottom 10 percent) olfactory test scores were followed over time to determine who developed Parkinson's disease.

Two years into the study, four of the 40 relatives (10 percent) with the worst olfactory test scores developed clinical Parkinson's disease, whereas none of the 38 relatives with the best test scores did. By five years, the incidence among relatives rose to 12.5 percent. Only those from the poor-smelling group evidenced decreased imaging of the dopamine transporter and, in some cases, such decreases were evident at baseline, suggesting that subclinical disease had already begun. These data suggested that the risk of developing Parkinson's disease in the presence of hyposmia may be as high as 22 percent in first-degree relatives.[35]

Numerous other studies support the view that olfactory testing may be helpful in predicting the development of cognitive and motor disorders.[24] Indeed, olfactory dysfunction rivals and even exceeds the sensitivity of a number of other biomarkers in such prediction, including SPECT imaging of the dopamine transporter,[36] as previously described. Currently, novel methods are being developed to enhance the predictive power of olfactory tests. For example, spraying the inside of the nose with atropine, a drug that accentuates cognitive dysfunction in patients with Alzheimer's disease via its anticholinergic effects, may induce a greater degree of smell loss in individuals who are at risk for future dementia, in effect "unmasking" the disease.[37]

Although quantitative smell testing is rarely performed by most physicians, it has been found to be useful in not only detecting, but also discriminating among, a number of neurodegenerative diseases. For example, a disorder called progressive supranuclear palsy (PSP) is often misdiagnosed as Parkinson's disease. Unlike Parkinson's, however, patients with PSP have a relatively normal sense of smell.[38] Thus, a neurologist whose evaluation of such a person is inconclusive can use smell testing to aid in making the correct diagnosis. The same is true for distinguishing between depression and Alzheimer's disease, with depressed patients showing little evidence of smell loss.[39] In a position paper, the quality standards committee of the American Academy of Neurology has in fact recommended that olfactory testing be considered to differentiate Parkinson's disease from PSP and corticobasal degeneration, another neurological disorder.[40]

Keeping Your Sense of Smell Sharp

How can each of us maintain a healthy sense of smell? All through life, viruses, air pollutants, and other environmental toxins cumulatively damage receptor cells In the uppermost regions of our noses. Such cumulative damage is not manifested until later in life, when few cells are left to provide a normal sense of smell. Hence, the degree to which we can minimize exposure to such agents, such as having good air filters in our homes and washing our hands regularly, can go a long way in preventing damage.

Maintaining good sleep patterns also protects smell function. Obstructive sleep apnea, in particular, has been associated with smell loss,[41] which may be attenuated by the use of a CPAP (continuous positive airway pressure) machine at night.[42] Importantly, daily exercise has been shown to help avert age-related decrements in the ability to smell.[43] It may even help preserve olfaction in persons with Parkinson's disease.[44]

Aside from maintaining good health via exercise, sleep, and avoidance of pathogens, a healthy diet and certain dietary interventions may also help. Gopinath and associates at the University of Sydney compared the odor identification ability of 1331 persons who differed in terms of their dietary intake of nuts, fish, butter, and margarine.[45] Those with the most nut and fish consumption had a reduced likelihood of olfactory loss, possibly because of the anti-inflammatory properties of these foods' constituents. Similarly, Richard Stevenson and his colleagues at Macquarie University in Sydney found that persons who ate a Western style diet high in saturated fat and sugar exhibited poorer odor identification test scores than those whose diets were low in these nutrients.[46]

Conversely, an impaired sense of smell may lead to dietary changes that ultimately impact health, as evidenced by a study at Yale University that assessed dietary habits in 80 elderly women, 37 of whom had olfactory dysfunction.[47] The latter exhibited a nutrient intake profile that increases the chances for cardiac disease, including a higher intake of sweets and a lower preference for citrus fruits.

A Tie to Coffee

My own colleagues and I showed, in 2007, that the lifetime history of coffee drinking was positively associated with UPSIT scores in relatives of patients with Parkinson's disease.[48] Thus, after correcting for age, gender, and tobacco use, the mean UPSIT score for those who drank less than one cup a day was 30.4 (out of a possible 40); for those who drank one cup a day it was 32.6, while for those who drank two to three cups a day or four or more cups a day the mean scores were 33.1 and 34.1, respectively. This pattern was statistically significant ($p < 0.009$) and was stronger in men than in women. Whether the same relationship may be found in the general population has yet to be determined.

Looking Forward

What does the future hold? In Japan, a major pharmaceutical company has begun an initiative to educate physicians about the importance of this primary sensory system and to distribute olfactory tests via pharmacies to promote better understanding of the health of the Japanese elderly population. In light of recent discoveries, it is inevitable that more and more physicians will begin to pay attention to the smell ability of their patients, and to routinely test this faculty.

Given that olfactory receptors are not just confined to the nose, future research will establish whether smell testing can provide information on the distribution of olfactory receptors elsewhere in the body and indicate whether such information may aid in the detection and treatment of some genetic-related diseases. More research on how odors influence mood and behavior is underway, and novel devices for adding odors to the environment are in development.

Just as we largely control our auditory environment, we may soon be able to engineer our olfactory environment in a much more sophisticated way, to alter our states of mind and improve our mental health. The future of olfactory research seems on target to fulfill the prescient assessment of Alexander Graham Bell in a 1927 *Scientific Monthly* article: "Odors are becoming more and more important in the world of scientific experiments and in medicine—and the need for more knowledge of odors will bring more knowledge, as surely as the sun shines."

9

The Illusion of the Perfect Brain Enhancer

By Emiliano Santarnecchi, Ph.D., and
Alvaro Pascual-Leone, M.D., Ph.D.

Emiliano Santarnecchi, Ph.D., is an instructor of neurology and co-director of the CME course in Transcranial Current Stimulation (tCS) at Harvard Medical School and a clinical research scientist at the Berenson-Allen Center for Non-invasive Brain Stimulation at the Beth Israel Medical Deaconess Center. His research involves Non-Invasive Brain Stimulation (NIBS), electrophysiology, and neuroimaging methods. He is currently testing the application of transcranial alternating current stimulation (tACS) to induce long-lasting changes in brain oscillations, which might translate into therapeutic options for patients with Alzheimer's disease Santarnecchi obtained his Ph.D. in applied neurological sciences at the University of Siena School of Medicine in Italy.

Alvaro Pascual-Leone, M.D., Ph.D., is professor of neurology and an associate dean for clinical and translational research at Harvard Medical School. He is chief for the Division of Cognitive Neurology and the director of the Berenson-Allen Center for Noninvasive Brain Stimulation at Beth Israel Deaconess Medical Center. Pascual-Leone is a practicing cognitive neurologist and researches the mechanisms that control brain plasticity across the life span. He is considered a world leader in the field of noninvasive brain stimulation, where his contributions span from technology development through basic neurobiologic insights from animal studies and modeling approaches, to human proof-of-principle and multicenter clinical trials. Pascual-Leone obtained an M.D. and a Ph.D. in neurophysiology in 1984 and 1985 respectively, both from the Faculty of Medicine of Freiburg University in Germany. He also trained at the University of Minnesota, the US National Institutes of Health, and the Cajal Institute of the Spanish Research Council.

Editor's Note: Many questions loom over transcranial direct current stimulation (tDCS), a non-invasive form of neurostimulation in which constant, low current is delivered directly to areas of the brain using small electrodes. It was first established in neuroscience research in the 1950s and 60s, but has seen rapid growth, particularly in the last five years. Originally developed to help patients with brain injuries such as strokes, tDCS is now also used to enhance language and mathematical ability, attention span, problem solving, memory, coordination, and even gaming skills. The authors examine its potential and pitfalls.

WHO DOESN'T WANT TO GET SMARTER? Who doesn't want to improve their grades, make better decisions, perform at work with greater success, be more skillful at a sport, or just excel at playing videogames?

Of course, there are ways to pursue such goals, but they require significant time, effort, dedication, and practice with no assurance of results. Any skill, from the simple ability to grab a cup of coffee to the complex challenge of piloting a fighter jet, depends on an intricate net of brain connections, shaped over many years of learning and consolidation processes. (The Spanish neuroscientist and Nobel Prize winner, Santiago Ramón y Cajal, was the first to make this point based on his notions about neuroplasticity, the brain's capacity to change and adapt.)

But our society has made us impatient, entitled, and unwilling to sacrifice and work hard. Most of us wish upon a star and hope for a miracle—or, at a minimum, a way to shorten the time and reduce the effort needed. Even if we weren't able to learn faster or more efficiently, perhaps most of us would settle for some help in improving on whatever we are trying to accomplish.

Unfortunately, not everyone is equally talented or fit for every job, sport, or academic career. For some people, it takes much longer than for others to gain a given skill, and many will never get there at all. All brains are not created equal, in other words. Current studies are in fact working to identify differences in brain structures and functions—"brain finger-prints"—that might predict an individual's response to a given intervention and cognitive training protocol, or capacity to acquire a new skill. The emerging appreciation of such differences, along with notions of genetic determinism, make it particularly tempting to seek ways to escape biological constraints, modify one's baseline talents, and become able to acquire capacities one could otherwise only dream of: to play soccer like Lionel Messi or tennis like Roger Federer, run like Usain Bolt, paint like Pablo Picasso, or sing like Luciano Pavarotti.

The Two Sides of Brain Stimulation

Over the last decade, a new class of "brain augmentation" products has entered the market, under the label of transcranial direct current stimulation, or tDCS. This neuromodulatory technique is one of several non-invasive brain stimulation (NIBS) methods that have been the subject of increasing neurobiological research, and have promising applications for various neuropsychiatric conditions.[12] But non-medical, largely experimentally untested applications are also being embraced—and raising concern.

tDCS has roots in 200 year-old experiments by Luigi Galvani, a pioneer of bioelectromagnetics, and his nephew, Giovanni Aldini.[3] It is thought to exert its effects through brain polarization: low-voltage electrical currents supposedly induce shifts in the concentration of free floating charged particles in brain tissue, making neurons less or more excitable depending on the polarity of the electrical field.[4] Compared to the complex hardware required for transcranial magnetic stimulation (TMS)5—the most established NIBS technique, for which several devices have been cleared by the Food and Drug Administration (FDA) to treat medication-resistant depression—tDCS stimulators are quite simple, the equivalent of a nine-volt battery and a simple electronic circuit connected to two or more surface electrodes.

Given the device's low cost, ease of application, and apparent safety across research studies, it is easy to understand why the use of tDCS outside laboratory walls has grown exponentially in the last few years. It has been marketed to increase cognitive abilities, including memory and abstract reasoning, enhance mood and energy, and even improve video-gaming skills and physical performance. Among athletes who reportedly have used tDCS are competitors at the Olympic Games in Brazil[6] and players for the Golden State Warriors of the NBA.[7] Professional video gamers were recently reported to have used tDCS in preparation for a $1 million competitive event, and online forums and blogs are full of do-it-yourself (DIY) guides for building devices that promise to make you smarter, faster, and more efficient.[8]

Such flamboyant non-medical applications are not without precedent: In 1804, Giovanni Aldini and several followers became celebrities who performed public demonstrations of brain "Faradizing" for entertainment at parties and public events, and boasted it might be even possible to revive the dead with the technique. They, however, lacked such 21st century refinements as Bluetooth capabilities that make some portable tDCS devices as ready as everyday gizmos.

Obviously, the reality of tDCS effects and those of similar devices based on different principles (e.g. ultrasound stimulation[9]) is more complicated, and we ought to learn from history and neuroscience, and avoid making mistakes that can be costly for the individual and the progress of knowledge.

Unique Neuroscience and Clinical Tools

NIBS provides unique tools to understand and modify brain function. These include transcranial magnetic stimulation (TMS), which induces electric current in the brain via powerful, brief magnetic pulses created by a coil held over the subject's head;[10] and variants of transcranial current stimulation (tCS)—direct or alternating current stimulation (tDCS, tACS) and random noise stimulation (tRNS)—which deliver low-voltage current through surface electrodes attached to the subject's head.[11,12] Temporally interfering electric stimulation (TI) promises to allow selective stimulation of deep brain structures. Transcranial infrared light stimulation modulates neuronal activity and brain oscillatory activity, while low-intensity focused ultrasound pulsations can induce neuronal excitation and inhibition without anatomical damage and with exquisite spatial precision, including the ability to selectively target white matter tracts.[13,14]

Such techniques are being adopted in fundamental, translational, and clinical brain research (Figure 1, adapted from[15]), and offer promise as diagnostic and therapeutic interventions for diverse diagnoses, symptoms, and disabilities. The field has been rapidly expanding, spurred partly by advances in computational, electronic, biological, and brain imaging technologies. While deeper understanding of the mechanisms of action of NIBS techniques is needed, studies using them can offer valuable insights into core principles of brain function and brain-behavior interactions. NIBS affects distributed brain networks, and can modulate brain activity in a controlled manner.[16,17] By inducing transient changes in brain activity, it can be deployed to experimentally test psychological and cognitive theoretical models, and to evaluate cause and effect relationships between a behavior (e.g., cognitive process) and activity in specific brain regions.

In the first demonstration of its potential, Pascual-Leone and colleagues used repetitive TMS (rTMS) to transiently induce speech arrest in healthy humans.[18] This work indicated that rTMS and similar brain perturbation approaches could be valuable to study the neurobiological substrates of human brain disorders, shedding light into the basis of neurological and psychiatric brain dysfunctions through "virtual lesions."[10]

More broadly, NIBS has offered insights into the neural substrates of mood and emotion,[19] decision making,[20,21] and moral judgement.[22] Combined with brain imaging and neurophysiologic methods, it can be used to characterize spatio-temporal properties of neural networks. Combining TMS with brain imaging makes it possible to examine the effects of a controlled perturbation of specific brain functional networks,[23,24] as was recently done to determine the relationships between behaviors and specific patterns of multi-region activity in the context of associative memory.[25]

Beyond *spatially* distributed neural networks, brain function involves the dynamic, i.e., time-varying, function of networks of brain regions. NIBS methods can offer unique insights into such *temporal* dynamics, including oscillatory mechanisms and patterns. Combined with EEG (an electrophysiological monitoring method to record electrical activity of the brain), they can be used to assess the temporal and spatial connectivity of brain regions in different behavioral and disease states,[26,27] the frequency response of different brain regions to brief stimulation,[28] and the compensatory frequency and connectivity changes that occur with longer-lasting perturbations.[27,29] In addition, NIBS can directly modulate the spatially-localized oscillations involved in complex cognitive functions,[30,31] including those dysregulated in disease states.

Importantly, NIBS approaches are applicable across the lifespan and offer translatable biomarkers to bridge the challenging divide between animal models and human studies. The real-time integration of NIBS with EEG or magnetic resonance imaging (MRI) enables the study of brain effects of the induced perturbation and can provide unique insights into spatial-temporal properties of the system that can be related to cognitive or behavioral abilities, used to classify clinical phenotypes, or serve as biomarkers for genetic explorations. The application of the same methods of assessing brain function in human and animal models offer translational phenotypes to integrate model (animal) systems and human studies.

In addition, there are growing clinical applications of NIBS. Four different systems for TMS have been cleared by the FDA for treatment of medication-resistant depression, and Medicare as well as most health insurance companies now provides coverage for it. A home-based, self-delivery TMS system has been cleared by the FDA for treatment of migraine, and another has been approved for pre-surgical mapping of motor and language functions. Several companies are conducting pivotal studies aimed at regulatory clearance of TMS and tCS devices for other indications, including schizophrenia, Alzheimer's disease, epilepsy, developmental disabilities such as dyscalculia and autism, and recovery from stroke.

Meanwhile, a growing number of hospital-based clinical programs and private practice clinics have been established, offering TMS and tCS to patients with a range of medical diagnoses, some "on-label" given FDA approval, and others "off-label," on the basis of published evidence but lacking FDA endorsement. Several companies are offering direct-to-consumer NIBS devices and there is a rapidly growing movement of do-it-yourself (DIY) adopters. However, fundamental questions remain, emphasizing the need for well-trained scientists conducting rigorous, high-quality studies on the mechanisms of action of NIBS modalities, properly powered studies on their therapeutic efficacy, and careful evaluation of potential risks, complications, and adverse effects.

Beyond Neuroscience and Clinical Medicine

In the case of tDCS, since its re-discovery by two independent teams of scientists in Italy and Germany[12,32] around two decades ago, this modality has been increasingly adopted in clinical and experimental settings.[15] Evidence-based clinical efficacy in large clinical trials is still moderate to weak, but studies show some value for almost any neurological and psychiatric disorder that has been tested, both in combination with other treatments (e.g., psychotherapy for depression and anxiety, cognitive remediation for schizophrenia, physical therapy for stroke) or as a stand-alone intervention.

As to enhancement of normal cognitive function, tDCS—as well as tACS and tRNS—have been used in attempts to amplify basically any domain, from attention to perception, language acquisition to working memory, visuo-motor coordination to intelligence. Although findings in these areas have often been presented in the context of underpowered, poorly controlled studies,[15] meta-analysis of available data (for an example see [33]) seem to support the efficacy of tDCS in different cognitive domains, but suffer from the likelihood of publication bias in the primary studies. Certainly, increasing knowledge about individual differences in the response to stimulation, the role of genetic determinants, and potential long-term (positive and negative) effects on brain function, suggest that we may just be observing the tip of the tDCS research iceberg.

Outside laboratory walls, there are more than a dozen companies marketing devices for brain stimulation, ranging from DIY kits for assembling a basic device at home, to high-end products like the "Halo" brain stimulation headphones, advertised as a $700 professional aide for elite athletes.[34] Detailed guidelines on how to build a tDCS device using $20 worth of electronic components are easy to find online, and a Reddit forum on tDCS which opened in 2013,[35] reports around 10,000 visitors per month. What might have constituted an underground culture has grown at an unexpected rate, leading to warning messages by the scientific community.[36,37] The issue is clearly not just the risks of DIY stimulation, but also its ethical regulation, with experts suggesting the need of new bioethical models.[38,39–41] Before addressing this knotty issue, it is important to understand what is actually at stake when we start "playing" with the brain.

Beware of Complexity

A tDCS device may be easy to build and the stimulation may be simple to apply, but its effects on the brain are far from straightforward. They represent the interaction between the applied current and the targeted brain area's structure and function, and it is worth remembering that the brain is the most complex network known to humans.[42] Can the application of a weak electric field for around 20 minutes at an almost imperceptible intensity operate in a controlled, safe, predictable manner, only inducing beneficial effects? When things seem too good to be true, they are generally false.

In a recent open letter to the tDCS DIY community published in *Annals of Neurology*,[37] Wurzman and colleagues pointed out a number of issues surrounding such use: that the parts of brain stimulated by tDCS and effects induced are less deterministic than one may think; that functional gains may be associated with meaningful tradeoffs; and that tDCS-induced brain changes may be long-lasting and could lead to enduring complications and deleterious effects as yet unknown. Before engaging in DIY brain stimulation, it seems critical to be aware of many of the unanswered questions about the mechanisms of action and effects of tDCS.

Among these areas of uncertainty: first, even though one may aim to target a given brain region, the effects of tDCS are never limited to the area between the electrodes. Current flows in complex ways, depending on which tissues it encounters, can affect the function of structures along its path. In addition, stimulation affecting activity in one region will always modulate activity in others because the brain is organized in intricate networks of interconnected regions. Through such widespread effects, tDCS may alter brain functions in unintended ways.

Second, selectively influencing a given brain function with tDCS is challenging. For instance, the main target of studies using tDCS to increase working-memory, attention, abstract reasoning, and mood has been the dorsolateral prefrontal cortex (DLPFC), which is highly interconnected with much of the brain, including cortical, subcortical, and brainstem regions. Presumably, tCS effects can spread across these diverse networks inducing behavioral and cognitive effects that will not be monitored during such experiments. In this context, it is also important to note that tCS interacts with ongoing brain activity and can have a different effect on neurons that are active during stimulation compared to those that are not. Thus, the specifics of brain activity, including the expectations of cognitive or behavioral consequences, can modify tCS effects. Stimulation while one is reading a book, meditating, visually fixating on a crosshair, watching TV, doing arithmetic, sleeping, or playing video games could all cause different changes in one's brain circuitry. In fact, even activity before tCS is applied may change its impact. At this point, we do not know how much such variables can influence the effect of tCS, or what is the best activity to achieve a certain change in brain function or behavioral impact.

Third, because of the complex organization of the brain and its dynamically interacting networks, modifying activity in one network with tCS can change the activity in others in a way that we cannot reliably predict: enhancement of some cognitive abilities may come at the cost of others. tCS may, for example, improve the ability to perform a task dependent on a given brain network, but hurt another because of interactions between networks.[43] It is important to note that such potential tradeoffs are generally under-recognized, as most studies focus on just one or two tasks, and deleterious effects may develop over time and only become recognizable long after the stimulation.

Fourth, changes in brain activity resulting from tCS, whether intended or unintended, may last longer than anticipated. Because of processes of brain plasticity, tCS may initiate a cascade of phenomena that could lead to enduring and even self-perpetuating brain changes. Studies suggest that cognitive enhancements (as well as concurrent tradeoffs) may persist for over six months after stimulation.[44] Ongoing, regular application of tCS may effectively sustain desired effects, but also increase risks of lasting undesirable consequences. Furthermore, the possible risks of a cumulative dose over years or a lifetime have not been sufficiently studied. Put bluntly, we simply do not know what potential long-term consequences tCS may have.

Finally, it is critical to remember that each of us is different and that the effects of tCS vary greatly across individuals: the impact of 20 minutes of tCS at a given intensity on the brain of a 65-year-old former accountant is very likely not the same as on the brain of a 16-year-old high-school student. Reading the scientific literature with an untrained eye might lead to cherry-picking data in favor of the expected outcome, while neglecting the fact that, for example, up to 30 percent of experimental subjects may respond to the identical tCS setting with cortical excitability changes in the opposite direction from most subjects.

The results reported in scientific papers are averages across participants, but frequently fail to address the high individual variability in response to stimulation.[45] Most importantly, even under controlled experimental conditions when a consistent modulatory effect is achieved in all the subjects, variance in the magnitude of the behavioral effects at the individual level is still high,[46] possibly leading to unpredictable and potentially detrimental consequences.[47] Such variability is clearly not monitored by the DIY community, and the same often applies to the scientific literature, in which negative findings are less consistently reported and shared on the Web. To complicate things further, gender, age, neuroanatomy, spontaneous brain activity patterns, hormones, genetic polymorphisms, and even prior exposure to brain stimulation can all potentially affect the result of a tCS session.[46]

Despite the above uncertainties and risks, numerous scientific studies apply repeated sessions of tCS with the intent of causing lasting changes in brain function. They are, however, almost always performed in patients with neurological or psychiatric brain diseases with the goal of alleviating symptoms, in clinical trials that are carefully designed and regulated, provide detailed disclosure of risks, and obtain informed consent of participants. Application of tCS to healthy subjects poses very different risk-benefit considerations. To complicate matters, many direct-to-consumer devices leave decisions as to stimulation timing, intensity, and duration up to users who may, in some cases, apply it multiple times per day for many months while doing different things. It is important to consider that because the impact of such long courses of repeated sessions of tCS have not been assessed, even in laboratory settings, users are truly unable to assess the risks to which they may be exposing themselves in the pursuit of cognitive enhancement.

The Right to Do It Wrong

Despite all open questions and uncertainty of risk, the most fundamental consideration might be whether we should be free, if we so choose, to subject our brains to whatever type of electrical stimulation we like. We would answer "yes," but it is a complex issue.[38,48] Are we facing a new type of "doping" which might require ad-hoc regulations, at least for its use in professional sport contexts? Is there enough evidence of risk to constitute a serious issue for the lay public, requiring some sort of medical monitoring when tCS is performed in any setting?

Perhaps, the response to these questions is "coffee." Even though coffee companies do not advertise that the effect of a shot of espresso has the spatial accuracy of tCS targeted to one specific brain region, the effects of coffee on cognitive functioning are indisputable.[49] So, why not regulate coffee intake? We also know that some studies found excessive coffee intake might be detrimental for general health,[50] but agree that self-regulation of coffee consumption is acceptable. On the other hand, one might say that if different types of coffee with selective effects on memory, attention, and logical reasoning were available, the same principle would no longer apply and we would have to discuss whether athletes should be allowed to take the espresso affecting reaction times while being free to take the memory-enhancing one.

So, is the difference simply that coffee's generalized, non-specific effect on brain activity makes it acceptable, while the specificity of tCS (it can—theoretically—target a given brain region, and directly affect brain physiology and associated cognitive function), represents "doping?" A lot of things affect brain physiology: talking with a friend about your troubles might make you change your mind about what to do tomorrow morning, because a series of chemical reactions in the brain strengthen and weaken different sets of connections, allowing a new thought to emerge to consciousness. The same applies when learning a new language or solving a puzzle.

The question is how long the same processes might take if induced by "natural" means (e.g., a conversation or 100 hours spent on a language-acquisition app on the smartphone) or if facilitated by making brain plasticity and learning processes more efficient through brain stimulation.

Any advancement that might help achieve a personal goal in the context of a fair competition (e.g., where everyone has access to and can use tCS) should be allowed, as everyone is permitted to use the latest cutting-edge swimsuit, the best vitamin supplement to recover after a soccer training session, or the lightest carbon-fiber bicycle on the market. The ethical principles of autonomy, equality, justice, and universalism apply. However, utilitarianism is a real issue because of the gaps in our knowledge of the potential consequences of long term use of tCS. It might well be worth asking what could be lost in the long term, rather than simply what might be gained in the short one.

10

The First Neuroethics Meeting: Then and Now

Essays by Jonathan D. Moreno, Patricia Smith Churchland, and Kenneth F. Schaffner.

Podcast transcript with Steven E. Hyman, M.D.

Editor's Note: It wasn't until 2002 that more than 150 neuroscientists, bioethicists, doctors of psychiatry and psychology, philosophers, and professors of law and public policy came together to chart the boundaries, define the issues, and raise some of the ethical implications tied to advances in brain research. On the 15th anniversary of the Neuroethics: Mapping the Field conference in San Francisco, we asked three of the original speakers to reflect on how far the neuroethics field has come in 15 years—and where the field may be going in the next 15.

How I Became a "Neuroethicist"

By Jonathan D. Moreno, Ph.D.

Jonathan D. Moreno, Ph.D., is the David and Lyn Silfen Professor of Ethics at the University of Pennsylvania. His most recent book, *Impromptu Man: J.L. Moreno and the Origins of Psychodrama, Encounter Culture, and the Social Network*, is about the life and times of his father the psychiatrist J.L. Moreno. He is also the author of *The Body Politic: The Battle Over Science in America* and *Mind Wars: Brain Science and the Military in the 21st Century*. In 2008-09 Moreno served as a member of President Barack Obama's transition team. His work has been cited by Al Gore and was used in the development of the screenplay for *The Bourne Legacy*. Moreno received his Ph.D. in philosophy from Washington University in St. Louis, was an Andrew W. Mellon post-doctoral fellow, holds an honorary doctorate from Hofstra University, and is a recipient of the Benjamin Rush Medal from the College of William and Mary Law School and the Dr. Jean Mayer Award for Global Citizenship from Tufts University.

THE NIGHT BEFORE THE NOW FAMOUS "Mapping the Field" conference in 2002 there was a dinner for the speakers. As I made my way to the restaurant, I wasn't sure what intrigued me more: the challenge of developing a presentation about neuroscience questions I'd never thought much about, or meeting the journalist, author, and our host Bill Safire (now deceased). In the event, the two became intertwined.

The neologism "neuroethics" was widely attributed to Safire, who had used the word in a *New York Times Magazine* column. As my turn to introduce myself came around, I rather nervously decided to cast caution to the wind. I said that I was a faculty member at the University of Virginia who was shocked to learn that Thomas Jefferson had not coined the term neuroethics. The hearty laugh I drew from the host, as well as his kind congratulations later for coming up with a good line, made the whole trip worthwhile.

And that was a good thing, because I consider my contribution to the panel that opened the meeting the next morning to be one of my less memorable efforts. Yet as one smart talk led to another, I began to appreciate the richness of this intersection of ethics and neuroscience. During the last session, an open discussion among the more than 100 people present, it suddenly dawned on me that one aspect of the new neuroethics had gone unmentioned. Two years earlier, I had published a book on the history of human experiments for national security purposes (*Undue Risk*), and was preparing an anthology on bioethics and biological weapons (*In the Wake of Terror*). Yet it even took me the better part of the day to realize that none of the speakers had considered why and how national security agencies could be interested in modern neuroscience, or the unique ethical issues that would flow from that interest. When I finally made a comment to that effect, I had the distinct impression that everyone in the room thought that I was from Mars. Except, that is, for one British gentleman who smiled at me with a knowing look. (I wish I knew who that was; if you're out there, please let me know!)

The seed that was planted in San Francisco took the form of the title of an article that I thought would be fun to write: "DARPA (Defense Advanced Research Projects Agency) on Your Mind." The phrase was rattling around in my head when then Dana Foundation editor Jane Nevins asked me if I would write something for *Cerebrum*. I don't think it took me more than three days to write the article. A few weeks later Jane passed on a question from Safire: Could I write a book for the Dana Press? I readily agreed (though in truth I had my doubts), and the result was *Mind Wars*. Of all the work I've done, I'm confident I will never be identified with anything more than I am with that book.

Since 2006, when *Mind Wars* was published, neuroethics has blossomed and the range of issues related to the book has expanded dramatically: brain–computer interfaces, neural nets, cognitive enhancement, external neural stimulation, autonomous weapons, interrogation, etc. A Japanese translation of the book soon followed, and a Chinese translation of the 2012 update (by the People's Military Medical Press, no less) is on the way.

Back when I wrote *Mind Wars*, I worried that my fascination with this offbeat topic would taint me as paranoid; 15 years ago the possibilities for "reading" and modifying the brain seemed far more distant than they do today. To my surprise I soon became a favorite source for journalists and the "go to" ethicist for various panels on neuroscience and national security, including, currently, a National Academies committee on behavioral science for the intelligence community. I have contributed vastly less to these endeavors than I have gained—among other things, the wonderful opportunity to bring new information and ideas into my classes at the University of Pennsylvania. Last spring I offered what must be the first undergraduate course on bioethics and national security, as usual with a good dose of *Mind Wars* thrown in.

All this I owe to Safire, Nevins, and the Dana Foundation.

The Brains Behind Morality

By Patricia Smith Churchland

Patricia Smith Churchland is a professor emerita of Philosophy at the University of California, San Diego, and an adjunct professor at the Salk Institute. Her research focuses on the interface between neuroscience and philosophy. She is author of the pioneering book, *Neurophilosophy*, and co-author with T. J. Sejnowski of *The Computational Brain*. Her current work focuses on morality and the social brain; *Braintrust: What Neuroscience tells us about Morality. Touching a Nerve* portrays how to get comfortable with this fact: I am what I am because my brain is as it is. She has been president of the American Philosophical Association and the Society for Philosophy and Psychology, and won a MacArthur Prize in 1991, the Rossi Prize for neuroscience in 2008, and the Prose Prize for science for the book, *Braintrust*. She was chair of the philosophy department at the University of California San Diego from 2000-2007. Churchland received a Bachelor of Philosophy from Oxford University, a Master of Arts degree from the University of Pittsburgh, and a Bachelor of Arts from the University of British Columbia.

HAVE WE LEARNED MORE ABOUT THE BRAIN basis for moral behavior in the years since that memorable neuroethics meeting in 2002? Yes, and remarkably so. Two branches of brain research that are especially relevant to this understanding have blossomed. First, social neuroscience is revealing the underpinnings of social bonding and why we trust and care about family and friends. This is the fundamental platform for morality. Second, research on the wiring that supports reinforcement learning is revealing how we acquire norms and values, along with the powerful feelings that accompany them. These are the social behaviors that take shape on the platform. What remains poorly understood is social problem solving—how social norms emerge or are modified in response to ecological and other pressures. These are the social institutions that give direction and predictability in a culture.

Linking Social Bonding and Morality

Moral values originate, not in the gods we invent, not in some magical "pure reason" detached from all feelings, but in the neurobiology of attachment. Bonding is powerful in mammals and birds, where the young are born immature and helpless. During their evolution, the brains of mammals and birds were rewired to ensure that the mother, and in some species also the father, bonded with and cared for their infants until they were independent. Caring for others, more generally, is a disposition that is expressed in a host of behavioral ways—provisioning, defending, cuddling, grooming, and being together. As a species adapts to its environment, these "ways" become established as values and norms.

Oxytocin is essential in the process of bonding. The mother mammal, whose brain is awash in oxytocin at the birth of her offspring, feels pain when separated from them, and relief when the babies are together with her. As babies are suckled and cuddled, their brains are awash in oxytocin, and they bond too, feeling distress when separated from the mother, and pleasure when close. Thus, stress hormones come into play as well. It is as though the ambit of self-care expands to embrace those to whom we are bonded. This seems to be the roots of all social glue in mammals and birds.

The suite of neurochemicals involved in bonding is not limited to oxytocin and stress hormones, but includes also the endogenous opioids and cannabinoids, which are associated with feelings of pleasure. Then there are galanin and vasopressin, which regulate aggression; and dopamine, the lynchpin of the reward system, which is essential for learning about one's social world. It is through the interplay of these compounds that we discover what is acceptable and what is not, which is nasty and who is nice, and how to get on in the social world. Anatomically, the ancient structures of the reward system, such as the hypothalamus and basal ganglia, are tightly linked to the neocortex, that amazing structure that gives mammals a level of flexibility not seen in social insects whose behavior is under tight genetic control.

Oxytocin research suggests that its roles are complex, not only in guiding attention to socially relevant goings-on and, probably, orchestrating the feelings accompanying bonding, but also in lowering levels of anxiety. In a first approximation of the process, as oxytocin levels rise, stress hormones fall; and with the increase in oxytocin levels come feelings associated with safety, friendliness, and sociality. This means that when I feel you are a friend, my stress hormones decrease and I can relax—I trust you and need not feel the vigilant anxiety that you may hurt me. We can cooperate; we can count on each other. Mammals that are bonded to each other are apt to console one another, want to hang out together, and perhaps share food and defend each other.

The hypothesis, briefly, is that the platform for morality is attachment and bonding, which in some species, extends to mates, kin, friends, and possibly strangers. In part I am drawing on a metaphor from computer engineering, where the platform—a computer's architecture and operating system— allows the software to run. Here, the neurobiology of attachment and bonding provides a motivational and emotional substructure that allows the scaffolding of social practices, moral inhibitions, and obligations to find expression. If mammals did not feel the powerful need to belong and be included, did not care about the well-being of offspring, kin and kith, then moral responsibility and moral concern could not take hold.

Learning Norms and Values

The brains of all mammals and birds are powerfully organized to construct and model their physical and social environments through learning. Learning takes off with a vengeance at birth, especially, but not only, in the cortex. Approval is highly rewarding, disapproval is the opposite. Here, too, evolutionarily ancient subcortical structures are crucial: the hippocampus and the basal ganglia provide the platform for reward learning. Links to the frontal cortex add sophistication and complexity in self-control and in sizing up a social complex situation, in planning and in evaluating consequences.

Is Morality Hard-Wired?

"Hard-wired" is an expression that makes me think of my toaster. It is "hard-wired" to toast bread and bagels. Period. But the mammalian nervous system is remarkable in that its genes turn off and on owing to environmental interactions, enabling changes in the very structure of its wiring. Brains adapt to local circumstances, in both the physical and the social domains.

Had I been born 250,000 years ago, I would have learned, as I grew up, to fit into a local social and physical environment very different from what we face today. With my acquisition of social knowledge managed by my basal ganglia and its reinforcement wiring, I might have found very hairy, smelly men exceptionally attractive, for example. (This does not mean that human brains are infinitely malleable—I can never learn to play hockey like Wayne Gretzky or sense odors like my dog Duff.) We might say my brain is "soft-wired" for sociality. I was born with a capacity to imitate, to want to be with others, to dislike shunning and disapproval. In a typical social environment of family and clan, those dispositions flourish and become very strong. In an abusive environment, they are apt to develop atypically.

Different cultures do have different practices regarding many aspects of social life, such as when it is wrong to stare or to laugh, when it is a social misstep to offer help, or the conditions under which one must sacrifice family needs for the needs of others. Such pluralism regarding moral practices is an important aspect of what we cope with as humans living together in a highly interconnected world.

Plasticity and Moral Practices and Values

Variability is, of course, a hallmark of biology, and there is certainly variability among humans in their predispositions to sociality, common sense, and temperament. Contrast the loner gold prospector, whose level of sociability extends no further than his old mutt, with the kindergarten teacher who is intensely social and loves to be in the middle of her loving brood. (Such differences in attachment needs notwithstanding, the old prospector might like to come to a pub once a year and the teacher might appreciate the quiet of her solitary cup of tea at the end of the day.)

Exposure from birth to close social interactions means that that the brain is shaped by the world right off the bat. When those worlds differ between individuals—as between that of an Inuit living in the Arctic and an Aztec living in Mesoamerica, there will be corresponding differences in what brains learn. But despite all that variability, it still seems likely that the basic urge to be with others—needed in all mammalian species for parents to tend infants and for infants to want to stay close to parents—is deep and strongly conserved. It is the platform for complex forms of sociality that we find pleasurable as well as profitable.

This means that we are connected to one another in a very basic way. It has long been observed that various social skills we acquire as we grow up are typically "transportable" to other families, other clans, other cultures, with perhaps a bit of tweaking. The brains of strangers can quickly synchronize owing to common networks and neurohormones, suggesting a powerful connecting thread that links us all together.

But certain differences persist: even when people agree on facts, they may rationally disagree on the relative importance of certain goals— about the value of freedom versus prosperity, for example, or of commitment to family versus dedication to all. Even if, as I suggest, the deepest level of value—the social platform—is shared, we may not agree on the relative value of higher-level practices, because learned values supporting those practices may differ. As I see it, these stubborn moral dilemmas are best addressed in a highly traditional way—through discussion and conversation, with respect and understanding, and by knowing as much as possible about the relevant history of different practices. My feeling is that knowing more about neuroscience, for example, is not going to solve these kinds of evaluative differences, whereas institutions that incentivize mutual understanding may at least encourage peaceful progress.

Past and Present Considerations

By Kenneth F. Schaffner, M.D., Ph.D.

Kenneth F. Schaffner, M.D., Ph.D., is Distinguished University Professor of History and Philosophy of Science emeritus at the University of Pittsburgh. He taught previously at the University of Chicago (1965-1972) and The George Washington University (1991-2005), where he is University Professor of Medical Humanities emeritus. At Pitt he served as co-director of the Center for Medical Ethics (1986-1991), and was editor-in-chief of *Philosophy of Science* (1975-1980). His most recent book is *Behaving: What's Genetic and What's Not and Why Should We Care?* (Oxford University Press, 2016), and he is currently at work on a sequel: *Choosing: What Can We Learn about Choice and Flourishing from Behavioral Neurogenetics?* Schaffner holds a Ph.D. in philosophy from Columbia University and an M.D. from the University of Pittsburgh.

THE FIELD OF NEUROETHICS HAS FLOURISHED in the 15 years since the 2002 landmark conference, "Neuroethics: Mapping the Field," and its subsequent publication--an expansion that includes the founding of the International Neuroethics Society, two journals (*Neuroethics* and *AJOB Neuroscience*), several related websites, and a number of significant books and collections as well as a recent federal research grants program from the "BRAIN Initiative: Research on the Ethical Implications of Advancements in Neurotechnology and Brain Science." This vibrant evolution has been nourished by the extraordinary development of neuroscience and neuro-imaging and by the hopes and concerns generated by the prospect of new brain interventions.

At the 2002 conference, I proposed several distinctions related to the neuroethics questions of reductionism and free will and, in the subsequent 15 years, have further explored these issues. But more needs to be said and speculated on involving scientific and philosophical developments in the past 15 years—and for the next 15 as well.

We can begin with the "standard model," the "split-level view" or the "hierarchical model," all of which are partly dependent on a development of the well-known theory of Harry Frankfurt and Gerald Dworkin—two celebrated American philosophy professors—on free will and autonomy.

Frankfurt's view notes that humans have the capacity to entertain "second-order desires," which is analogous to an actor deliberating about choice and decision. When an individual acts, so that both first- and second-order desires agree, that individual exhibits "free action." Another way of looking at that agreement is that the action reflects the individual's "true self," and that the second-order desire is one in which the individual identifies and is happy to own.

Frankfurt's account has generated extensive criticism, but an elaboration of his approach has become a sort of "default position" in neuroethics, with the agreement of first and second order desires being supplemented with requirements of rationality and information sufficiency, as well as the absence of external pressures or excusing conditions such as mental illness. In the neuroethics literature, such action involving free will is frequently termed "agency."

Still, more examination is required, and this free-will perspective suggests that a deep account of a person's "self" might offer an approach to the intertwined notions of choice, freedom, control, and the fundamental value of autonomy. In my view, one might consider *self* in terms of what is stable over the long term—what "grounds" a person and what he or she is like. Their "self-identity" or "personal identity" or "who we are" is constituted by a temporarily evolving but roughly stable continuous narrative. From extensive psychological research as well as human genetics investigations, we may well best obtain a reasonable understanding, or at least a first attempt at such, by examining the nature of an individual's personality. The self, however, surely encompasses many other facets, a view developed originally by the 19th century American philosopher and psychologist William James, and modernized by Ulric Neisser, a German-born American psychologist and author of *Cognitive Philosophy* in 1967.

Ethics and Neuromodulation

In recent years, neuroethics concerns in the areas of choice, personality, and the self have been complicated by the development of brain neuromodulation techniques. Closely related issues involve personality alterations and the limitations of choice posed by such conditions as addiction and psychopathy, which may adversely affect both the individual and society and become entangled with legal questions. It is in the complex area related to neuromodulation that I envision significant neuroethics advances and challenges emerging.

Three neuromodulation interventions thus far have been shown to affect motor diseases, psychiatric disorders, executive processes, and personality traits. All are currently under investigation in research contexts and two are used clinically. Deep brain stimulation (DBS), in which electrodes are implanted deep in the brain and connected to a permanently implanted electrical source, is invasive, but also the most effective. Externally applied magnetic and electrical interventions—repetitive transcranial magnetic stimulation (rTMS) and transcranial direct current stimulation (tDCS)—are noninvasive. The latter techniques have had less consistent results than DBS but offer promise, and one (rTMS) has been approved for treatment of depression.

Questions surrounding DBS, which has proven quite useful in the treatment of Parkinson's disease, depression, and obsessive-compulsive disorder (OCD), have arisen with the appearance of marked and unusual side effects. One of the most striking examples in the literature involved a 60-year old OCD patient who developed a sudden and distinct musical preference for Johnny Cash following DBS, but who reverted to his usual, more eclectic tastes when not under neuromodulation. A more distressing case involved a 62-year old patient with severe Parkinson's disease that was alleviated with DBS, but who developed severe mania while stimulated, resulting in the ethical dilemma to the patient of choosing to be either bed-ridden or manic. He made the latter choice.

Ethical questions are also raised by the application of DBS, now in research trials, to the treatment of addiction behavior, and by the possibility (currently under pre-clinical investigation) of using it for personality disorders, including psychopathy or its more restrictive DSM-related construct, antisocial personality disorder. This type of intervention has been proposed for potential treatment of incarcerated but willing psychopathic individuals.

(Though it is not known exactly how DBS works, recent advances in optogenetics—a precision technique developed by Stanford University professor Karl Deisseroth in 2005, in which light-responsive proteins allow specific neurons to be turned on or off—may resolve these questions.)

DBS, moreover, can produce significant neuroanatomical circuit remodeling (as recently shown in a mouse model of depression), raising troublesome questions concerning reversibility. To resolve such questions, much more will need to be known about brain circuits and how their variations affect thought and behavior. Advances in this area are likely to come from the National Institute of Mental Health (NIMH) RDoC initiative, already under way.

For such reasons, neuromodulation interventions have already generated concerns and policy proposals in neuroethics, and will continue to do so with increasing urgency. Some philosophical and psychological developments offer assistance. One argument states that psychological and neuroscientific findings amplify our understanding of choice behavior but undercut the "standard model" cited above, showing it to be illusory. Others, however, see that something like this standard model can be retained, albeit with further refinement. Dartmouth College philosophy professor Adina L. Roskies has rightly suggested that "agency" in a DBS context—and of the type approximated in the standard model—needs much more multidimensional development to reflect its true complexities.

Recommended Philosophers

With these approaches in mind, neuroethicists should examine the extensive philosophical literature on the "self," such as found in books edited by Shaun Gallagher, and especially in philosophical analyses of self-identity. These inquiries need to be coupled with neuroscience advances as pioneered by neuroscientist- philosophers like Antonio Damasio and Georg

Northoff, in a combined top-down and bottom-up pincer strategy. Neurobiological uncertainty about the self—in particular, its anatomical location—makes this an especially challenging task. Commenting on localization, Gallagher has written that "there seems to be overwhelming evidence that the self is everywhere and nowhere in the brain."

Further, I believe they should incorporate the burgeoning psychological literature on the "true self" which is a kind of conscience or super-ego. This notion overlaps with the concept of the cluster of second-order desires in the standard model, and shares with that concept features both of subjectivity, normativity, and complexities of verifiability. Including a "true self" in this mix, however, with subjective and necessarily person-variable features, will be controversial. And perhaps, along with the notion of self-identity, even constitute a *diachronic* version of the much-discussed but still unsolved "hard problem" of what consciousness is, as initially formulated by philosopher David Chalmers.

This possibility that we will confront the "hard problem" of consciousness in neuroethics in this regard seems very likely, since first-order desires are perceived as first-person experiences, and second-order desires are similar, even if intertwined with memories and critical executive evaluations. Given its centrality in the issues of choice and free will and the standard model, and though difficult and understudied, including consideration of a "true self" in these analyses seems worthy of sustained exploration. It is these types of investigations into a deep account of a person's "self" that I believe will advance this significant set of interrelated neuroethical issues.

Down the Line

This, then, is my expectation: neuroethics will increasingly be concerned with ethical, psychological, psychiatric, and legal issues in the intertwined areas of choice, personality, the self, and consciousness as affected by continuing developments in brain neuromodulation techniques. Refinement of these techniques, especially DBS, is likely to come from integrated optogenetic capacities that will provide better knowledge, specificity, and control. More generally, advances in the understanding of brain circuits and higher-level functions including consciousness, arising out of research such as the NIMH's RDoC initiative, will foster progress in the field.

References to specific citations supporting most of the neuroethics advances mentioned are available from the author by writing him at kfs@pitt.edu.

Transcript of Cerebrum Podcast: Neuroethics Pioneer Steve Hyman

Guest: **Steven E. Hyman**, M.D., is the director of the Stanley Center for Psychiatric Research at the Broad Institute, the Harvard University Distinguished Service Professor of Stem Cell and Regenerative Biology, and a member of the Dana Foundation Board of Directors. Hyman joined the Broad after a decade of service as provost of Harvard University, where, as Harvard's chief academic officer, he focused on the development of collaborative scientific initiatives. From 1996 to 2001, he served as director of the U.S. National Institute of Mental Health (NIMH). Hyman is the editor of the *Annual Review of Neuroscience* and the founding president of the International Neuroethics Society. Prior to his government service he was the first faculty director of Harvard University›s interdisciplinary Mind, Brain, and Behavior Initiative, where he studied the control of neural gene expression by neurotransmitters with the goal of understanding mechanisms that regulate emotion and motivation in health and illness.

Bill Glovin: Today on the *Cerebrum* podcast, we have a very special guest under very special circumstances. I'm *Cerebrum* editor, Bill Glovin, and I'd like to welcome in Dr. Steve Hyman. Why are these special circumstances? Well, for one, I generally conduct these podcasts on the phone but today, I'm sitting across from Steve at the Marriott Marquis in Washington DC, which is connected to the convention center that is hosting this year's Society for Neuroscience (SfN) Conference.

The other reason it's a little bit different than usual is we usually feature the author of one of our most recent *Cerebrum* articles. In this case, we are making an exception because Steve is one of the founders of the Neuroethics Society, which meets each year a few days before the SfN conference.

Some of our conversation will be linked to our most recent *Cerebrum* article, which is titled "The First Neuroethics Meeting: Then and Now," which consists of essays by three of the original attendees of the Mapping the Field Conference in 2002 in San Francisco. Steve was also one of the original contributors at the Mapping the Field Conference, which is celebrating its 15th anniversary this year. That conference is unofficially, or maybe officially, considered the first time the neuroscience world came together to discuss neuroethics, which has just exploded in recent years as an integral part of brain research.

The last few days, Steve has been busy at the society meeting, which was filled with all kinds of fascinating lectures and panel discussions and jam-packed with attendees. When Steve isn't busy at these kinds of meetings, he's director of the Stanley Center for Psychiatric Research at Broad Institute of MIT and Harvard. He's also Harvard University distinguished service professor of stem-cell and regenerative biology, and has served as provost of Harvard, as the university's chief academic officer, and as president of SfN.

Steve, instead of me rambling on about the Mapping the Field Conference, I'd rather have you tell me your memories of how it all came together.

Steven Hyman: In many ways, the inspiration for bringing it together came from Bill Safire (former *New York Times* columnist and chairman of the Dana Foundation), who had been thinking about the rapid advances in technology to investigate and, ultimately, to influence the workings of the human brain. He had an insight that, what he came to call neuroethics, would extend far beyond the kind of important but conventional bioethics that many people were already involved in.

What I mean by conventional bioethics is the platform for research, how we get informed consent, how we make sure that we protect human research subjects, and the like. What emerged was a far more interdisciplinary field with concerns around things like brain privacy; once we were able to examine the brain, perhaps in new ways, or human identity, meaning, did we have interventions in the brain that would alter human personality or human sense of self, issues like memory editing and then other issues like cognitive enhancement. I don't want to go on with a laundry list but you can see how fundamentally important these advances are to our sense of what it means to be human.

Bill Glovin: It seems that neuroscience had established itself as an incredibly important part of medicine long before 2002. Why had this Mapping the Field Conference not taken place 10 years or even 20 years before?

Steven Hyman: I think that's a really important question. Neuroscience, in its early decades, was extremely successful but was largely focused on non-human models that were allowing us to understand how neurons talk to each other, how memories were stored, but much of the important work was done in model organisms ranging from the sea slug, Aplysia, which has a simple and tractable nervous system, to rodents. What happened, progressively however, is we began to understand human neuroscience, number one, and number two, to develop very powerful technologies, both to examine the living human brain such as magnetic resonance imaging, which allows both studies of the structure and function of the human brain, but also to develop new interventions directly into the human brain.

Let's take the example of deep brain stimulation, which is now widely used, perhaps there are 200,000 people worldwide being treated with this, but this involves putting a depth electrode chronically into a person's brain and stimulating certain brain regions, with the greatest success being in treatment of movement disorders like Parkinson's disease and severe essential tremor. It became recognized that some people have side effects of such treatments that change their impulse control or aspects of their personality. I think Bill Safire recognized pretty early that we were entering a new era of the ability to understand and manipulate the human brain as the seed for our thoughts, our emotions, our self-control.

Bill Glovin: What was the takeaway from the conference and how did you get other leaders in the neuroscience world to keep the momentum going from that conference?

Steven Hyman: An important takeaway from the conference was the idea that we really did have to invent a new interdisciplinary approach. Many of the people at the meeting were practitioners of bioethics and they had thought for their whole lives about how do you get informed consent for research or for clinical care from somebody who was in cognitive decline or somebody with a mental illness. How do we protect such vulnerable people?

We recognized we had to supplement that approach with philosophers of mind, with lawyers who were thinking about under what circumstances might we begin to worry about somebody's moral responsibility or legal culpability. We had to include engineers and computer scientists because artificial intelligence was increasingly—even 15 years ago—making inroads in both understanding data about human beings but also mimicking and maybe improving upon some of the things that human beings don't do so well.

The big takeaway was this had to be a big tent and then the second thing is that it was really early days, that this would fall apart unless there was a sustained conversation among people who were anchored in difference disciplines, including, of course, neuroscience.

Bill Glovin: Were there any specific events in brain research that said, "We, as neuroscientists, need an organization?"

Steven Hyman: I think that there was no one. specific event, rather, there were a series of discoveries that, by the way, continue to this day and, while the International Society for Neuroethics (initially just called the Neuroethics Society), was founded 15 years ago, the membership has grown and the field has grown, both inside and outside the society as new discoveries are added that might influence the way we look at our brains and can influence our brains.

For example, recent research on memory editing, which is being conducted in order to perhaps treat people with severe post-traumatic stress disorder but, if one can begin to edit memories that are encoded under strong emotions, there are concerns as well about human identity and human narratives. After all, that's what Macbeth asked Lady Macbeth's doctor for "canst though not minister to a mind diseased." Anyone troubled by conscience might also want memory editing. These are the kinds of things that really demand a deep subject matter base but also ethical analysis.

Bill Glovin: And neuroethics has seemingly become its own academic discipline in just 15 years. Does that surprise you?

Steven Hyman: It does surprise me, actually. I would say it's emerging as a discipline so, at the International Neuroethics Society meeting yesterday, there were a very large number of young people, some of whom even were doing Ph.Ds in what might be called neuroethics, even if their universities didn't have a designated program yet, many post-doctoral fellows who were working with people who would call themselves neuroethicists. I think it is an emergent discipline. It is really driven by a series of stunning advances in neuroscience and especially in human neuroscience and the recognition that we better be thinking deeply about these things and where they're going because for every advance, there's also potential downside complexities that perhaps people have not yet imagined.

Bill Glovin: You mentioned brain stim, consciousness, and memory. Are there some other areas that might be part of the new frontier?

Steven Hyman: Well, there are whole host of areas. One very interesting area is gene therapy. There were just some approvals for blinding retinol diseases, retina being the most accessible part of the brain, but where one uses a virus that's been engineered to carry a genetic payload to correct something, and it won't be long before we start seeing gene therapy for a variety of brain disorders that initially will be of course very, very severe disorders.

Now, it's important, gene therapy doesn't change your germline, it's not passed on to your children, but it's a direct manipulation of the genetic material in the cells, in this case, of your brain. This could have all kinds of implications, the most important ones being treating dreadful diseases but again, we don't know what manipulating the brain is going to do to somebody's sense of self, their identity, their memory, their self-control?

Another area that is emerging is our advances in deep brain stimulation. To date, deep brain stimulation has involved putting an electrode into somebody's brain and then, the doctor working with the person who has the stimulator, comes up with the best frequency with which to stimulate, it might change over time but it's very much a feedback loop that involves patient report symptoms, observed symptoms, the doctor and metaphorically, the twiddling of dials. But that's pretty inefficient; so, what's called closed loop deep brain stimulation is now very close to implementation, where you can have electrodes that not only "write," that is stimulate, but also "read," that is, you might be stimulating one part of the brain, to let's say, release more dopamine, a key neurotransmitter involved in control of movement but also in motivation and other important functions, including deciding which memories might be important to encode, and then you have a second electrode somewhere else in the brain that is reading out the effects of having released that dopamine. Then there's an algorithm that feeds back to the stimulating electrode.

Let's imagine that somebody with closed loop deep brain stimulation or read/write electrodes that involve dopamine (and I selected dopamine on purpose because it also involves, when things are overdone, impulsive behavior). Let's say somebody then commits an illegal act; who's responsible? Is the person still a moral agent? Is it the algorithm that's at fault? In some ways, these are very similar problems to the problems of programming autonomous vehicles when they have to choose between two different, inevitable accidents, but this is actually within the human brain. I think the issues raised are profound and require, again, a new kind of interdisciplinary thinking.

Bill Glovin: Last month, Alvaro Pascual-Leone, a colleague of yours at Harvard, did a *Cerebrum* article for us. He spent a lot of time, in the article, talking about the neuroethics issues of it and how people are not using it properly, how it's being commercialized, so that is a great example of someone on the front lines really recognizing all the dangers.

Steven Hyman: I think Alvaro is most concerned, of course, not about these neurosurgically implanted invasive electrodes, which are still very much under the control of medical investigators but also increasingly clinical, but noninvasive brain stimulation apparatus that you can buy on the web or you can make it yourself in your garage. Often, people think that these things are, at worst, harmless and, at best, will advance their cognitive abilities, for example, or make them more alert. But in fact, we have no idea and there's some research to say that even if they work at the margins, perhaps you increase one function at the expense of another function. Not to be too much of a naysayer but I think the idea of exposing your brain to electrical current chronically, without much research, is probably not something I would recommend to a friend.

Bill Glovin: Sure. Since I brought up Alvaro, who are some of the other top thinkers in the neuroethics field? Can you recommend some works to listeners interested in neuroethics?

Steven Hyman: Right now, one of the really exciting things for the development of the field is that there is this global interlocking set of brain projects that are sponsored by government funded research agencies like the Nantional Institues of Health, and the National Science Foundation and the United States, but they're brain projects in China, Japan, Korea, Israel, and across the EU. Again, as these projects advance brain research, very explicitly in the United States, an ethical component was mandated, so there's now a neuroethics division of the NIH brain project, which brings me around to your question. One of the important leaders of that is a traditional bioethicist, a very good one, Christine Grady.

One of the founding neuroethicists, meaning somebody who was at the initial Dana Foundation meeting 15 years ago and part of the founding of the International Neuroethics Society, that's Hank Greeley, who really has rooted neuroscience and the law into neuroethics.

Judy Illes, who is just finishing her term as President of the International Neuroethics Society, has just produced a volume from Cambridge University Press on neuroethics; maybe some of the articles are a bit dense, but there is a lot of good background.

There have been no shortage of excellent neuroethics articles in *Cerebrum*.

Bill Glovin: Thank you. Yesterday I was in the audience when William Safire was posthumously awarded the Steven E. Hyman Award for distinguished service to the field of neuroethics. I've never had an award named for me, that seems pretty great. What about your own role? That's just a great honor itself.

Steven Hyman: Yeah, it was. I was really touched. On the other hand, I agree with you, it's very unusual. I worry that having an award named after me means that some people think I'm already dead.

I had actually been an undergraduate philosophy major and then I spent two years at the University of Cambridge doing history and the philosophy of science and was going to do a thesis in philosophy of science, when I decided no, you know what I really want to do is understand the biology underneath these issues that I'm so curious about. I took a circuitous route through medical school into the lab doing neuroscience but the initial questions that I brought with me, and actually I think a lot of neuroscientist are motivated by, which is: "Who am I? How do I think? How do I formulate an identity? How stable is it? How real are my memories? How are the encoded? What does that mean for me as a moral agent?" they lurked, always, inside my mind in spare moments when I wasn't pipetting something in the lab.

Then I had the opportunity to become the director of the National Institute of Mental Health and, while again, the cornerstone of the job was making sure that we were programmatically focused on funding the very best neuroscience, the very best psychiatric genetics, the best clinical trials; and an enormous host of ethical issues arose. They included traditional issues like: how do you do the best-informed consent in somebody who is acutely psychotic (if you want to be able to study that phase) but also issues of what happens long term to children who are treated with psychotropic drugs. On the one hand, you want to control symptoms that are keeping them from learning in school but, on the other hand, what are you doing to their brain development and how do we even get a handle on those tradeoffs, managing some very impairing, immediate concerns where we can see the harm and balancing that against some unknown, possible downstream consequences.

That's just one example but that rekindled my interest. I began writing about these issues, sort of as a hobby. I've written a number of articles about volitional control of behavior and agency in light of neuroscience, often using drug addiction, which is something I had worked on years ago, as an example. When the Dana Foundation crystallized this field, a group of us got together to think about how we would advance it and ultimately formed a society. I had this longstanding interest but I should say, my day job was being a neuroscientist, but I guess maybe it fell to me to be the first president and to be president, actually for too long, but while the society was forming because I'd had these administrative posts and I knew how to make things work and how to put together an organization and read a spreadsheet, things like that, not too elevated but important.

Bill Glovin: Getting back to the society for a minute, besides the annual meeting, what else does the society do, A, and B, how can people get involved if they're interested?

Steven Hyman: Well, I think there are number of ways to get involved, one is to look to the society's really excellent website, which you can find if you google "International Neuroethics Society" but there are lots of good writings that come from the society. One very important activity is the relationship to the *American Journal of Bioethics Neuroscience*, so there's the *American Journal of Bioethics* but it has this special neuroscience offshoot, that's edited by Paul Root Wolpe, who, again, is one of the founders of the society and a very big thinker. It is the official journal of the society, although there are other journals. It has a blog, which is very easy to follow, that is worth looking at.

Bill Glovin: So where do you see the field 15 years from now?

Steven Hyman: I'm actually quite optimistic. I think it will coalesce. I think there'll be some graduate programs and, again, I'm judging that based on the growing power of neuro technology, they're going to demand a lot of reflection in their application; the request from a number of neuroscience departments but also federal agencies to have a "neuroethicist" or people helping them think about these neuroethics issues, combined with the large number of young people who were at the Neuroethics Society meeting yesterday. I think of these bright, young students as the leading indicator that there's something there.

Bill Glovin: Anything else that I have failed to mention that you think is important to this?

Steven Hyman: I think, to come full circle, this is very much an interdisciplinary effort but I think we still have a long way to go in terms of these different disciplines really understanding each other. Very typically at our meetings or in journal articles, you can see polite but serious disagreements between the way criminal law would like to see human volitional control and the way neuroscientists see volitional control. This is an unsettled issue, right, where the legal system is based on the idea that, for the most part, we freely decide what to do and then act according to those decisions, and neuroscientists would, at a minimum, caveat that a great deal in the way that the criminal justice system would find very, very difficult to manage.

What I would foresee, without knowing the outcome of those discussions, as one example, is this very rich discussion across disciplines that's going to take a lot of twists and turns, and frankly, also has to be international because different systems of law will have different responses to the insides of neuroscience. It's going to be an exciting future.

Bill Glovin: An exciting future is always a great place to end. An enormous thanks, again, to Steve Hyman, who did just a masterful job of articulating the underpinnings of the neuroethics field. His enthusiasm is pretty contagious, I'd say.

You may have noticed some background and an echo during the last question, that's because people started coming into the meeting room we were in before we expected them to and we had to move out into the hall.

Let me just say that I failed to mention in the beginning of this podcast that Steve has also served as director at the National Institute of Mental Health, or NIMH as it's known in the field. Steve alluded to it at one point.

Once again, I'm Bill Glovin and all our content is brought to you by the Dana Foundation. You can access all or any of it at DANA.org, and this podcast on multiple platforms including iTunes, YouTube, Spotify, Stitcher and SoundCloud. See you next time.

11

Microglia: The Brain's First Responders

By Staci Bilbo, Ph.D., and Beth Stevens, Ph.D.

Staci Bilbo, Ph.D., is an associate professor of pediatrics and neuroscience at Harvard Medical School and the director of research for the Lurie Center for Autism at Massachusetts General Hospital for Children. Her research is broadly focused on the mechanisms by which the immune and endocrine systems interact with the brain to impact health and behavior. Bilbo was the recipient of the Robert Ader New Investigator Award from the Psychoneuroimmunology Research Society and the Frank Beach Young Investigator Award from the Society for Behavioral Neuroendocrinology in 2011. Bilbo received her B.A. in psychology and biology from the University of Texas at Austin and her Ph.D. in neuroendocrinology at Johns Hopkins University. She was on the faculty at Duke University from 2007-2016 before joining the faculty at Harvard in 2017.

Beth Stevens, Ph.D., is an associate professor at Harvard Medical School in the FM Kirby Neurobiology Center at Boston Children's Hospital and an institute member of the Broad Institute and Stanley Center for Neuropsychiatric Research. Her most recent work seeks to uncover the role that microglial cells play in neurodevelopmental and neuropsychiatric disorders. Stevens was named a MacArthur Fellow by the MacArthur Foundation in 2015, received the Presidential Early Career Award for Scientists and Engineers, and the Dana Foundation Award and Ellison Medical Foundation New Scholar in Aging Award. Stevens received her B.S. at Northeastern University, carried out her graduate research at the National Institutes of Health, and received her Ph.D. from University of Maryland. She completed her postdoctoral research at Stanford University with Ben Barres.

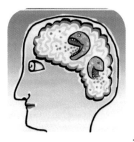

Editor's Note: New knowledge about microglia is so fresh that it's not even in the textbooks yet. Microglia are cells that help guide brain development and serve as its immune system helpers by gobbling up diseased or damaged cells and discarding cellular debris. Our authors believe that microglia might hold the key to understanding not just normal brain development, but also what causes Alzheimer's disease, Huntington's disease, autism, schizophrenia, and other intractable brain disorders.

EARLY IN THE 19TH CENTURY, the nervous system was believed to be a continuous network— essentially one giant cell with many spidery extensions bundled to form the brain and spinal cord. The discovery that nervous tissue, like any other bodily tissue, is composed of individual cells upended this theory, but the idea of interconnectedness persists.

Indeed, one of the most surprising findings in the neuroscience field in recent years is the degree of the nervous system's interconnection. We've learned that its cells are intertwined not only with each other but also with those of the immune system, and that the same immune cells that work in the body to repair damaged tissues and defend us from infections are also critical for normal brain development and function.[1,2] Some of these immune cells, called microglia, live permanently interspersed with neurons in the central nervous system and play crucial roles in nerve cell development, brain surveillance, and circuit sculpting.

In an article about microglia in *Biomedicine* in 2016, the author wrote that "scientists for years have ignored microglia and other glia cells in favor of neurons. Neurons that fire together allow us to think, breathe, and move. We see, hear, and feel using neurons, and we form memory and associations when the connections between different neurons strengthen at the junctions between them, known as synapses. Many neuroscientists argue that neurons create our very consciousness." However, what we know now is that neurons don't function very well, or at all, without their glial cell neighbors.

There is, in fact, perhaps no more dramatic a shift in focus in recent neuroscience than the ascent of these "other brain cells"—so dramatic in fact, that the knowledge has yet to seep into neuroscience textbooks and has only just begun to permeate the field. This knowledge, some of it described here, likely represents only the tip of the iceberg.

The Collective World of Glia

Microglia are the permanent resident immune cells of the brain and spinal cord, sharing many similarities with macrophages—the cells that destroy pathogens—outside the central nervous system. First impressions were underwhelming. In the 1800s, the pathologist Rudolf Virchow noted the presence of small round cells packing the spaces between neurons and named them "nervenkitt" or neuroglia," which can be translated to putty or glue. One variety of these cells, known as astrocytes, was defined in 1893.

Microglia themselves were first identified and characterized by Spanish neuroanatomists Nicolas Achucarro and Pio del Río-Hortega, both students of Santiago Ramon y Cajal, the undisputed "father of neuroscience," early in the 20th century.[4,5] Urban legend has it that del Rio-Hortega suggested that microglia looked like aliens from another realm—which is, metaphorically, not far off, given their origin in the fetal yolk sac[6] rather than the neural ectoderm from which all other brain cells develop. The relatively late entry of microglia into the neuroscience field a century ago may be in part responsible for the limited attention and understanding they have received. But since their origin was fully described seven years ago, the importance of microglia has gradually been recognized.

Microglia are distributed more or less uniformly throughout the adult brain, in both white and grey matter, but in varying densities, with the highest concentrations appearing in parts of the brain stem (the substantia nigra), parts of the reward circuit of the brain (the basal ganglia), and the hippocampus.[7] Each cell has a small cell body and numerous arms that extend throughout the surrounding tissue (Figure 1), maintaining distinct boundaries and rarely overlapping with the arms of a neighboring microg-lial cell.[8] Like police officers, these cells constantly survey their environment for trouble and are often the first responders to injury or disease. On their surface are a tremendous diversity of receptors for various threats, including bacterial, viral, and fungal pathogens, toxins, and xenobiotics, as well as noxious compounds released from dead or dying cells during traumatic brain injury, ischemia, and neurodegeneration.[9-12] Microglia from different brain regions are also somewhat heterogeneous, possessing a different collection of cell surface markers (sort of like little flags on the membrane that distinguish one cell from another), though the functional consequences of these differences are not yet fully understood.[13]

Upon detection of trouble, microglia mount specialized responses, destroying pathogens and calling for help from other cells via signaling molecules called cytokines. They organize the responses of surrounding cells to

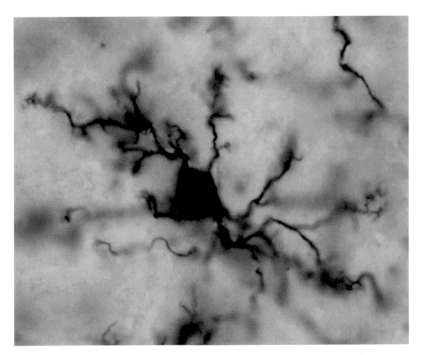

Figure 1. Microglia in the mature, healthy brain exhibit small cell bodies and multiple long, thin processes (arms) that they use to constantly scan and survey their local environments within brain tissue. Photo credit S. Bilbo.

alter neuron function, recruit additional immune cells, aid in tissue repair, or induce cell death.[8] Their constant communication with neighboring neurons and microglia ensure that each microglial cell is adequately placed and functioning at the right level of activity.[14]

Microglia Never Rest

It was traditionally assumed that microglia remained in a resting or quiescent state until mobilized by a threat, a transformation termed activation;[15] the cells retract their arms and adopt an amoeboid shape in which they can move spontaneously and actively.[16] In recent years, however, the notion of "resting" microglia was upended by a series of elegant experiments.[17-19] Using a green florescent protein (GFP) to color microglia and fancy two-photon microscopy to image them, researchers could watch these cells survey the brain through the thinned skulls of mice. Time-lapse videography revealed that while the bodies of cortical microglia remain relatively stationary, their arms are highly and spontaneously active, collectively surveying the entire brain every few hours.[19] These studies indicated for the first time that microglia are not simply "reactive" immune cells that mobilize following infection or injury, but active sentinels (Figure 2).

But why must microglia be so active if they are merely watching for threats? Several groups have argued that they play an essential role in monitoring synaptic activity as well.[14, 21-24] Synapses, the connections between neurons, are in effect the telephone wires of the brain, allowing these cells to electrically communicate with one another using their axons as transmitters and dendrites as the receivers. Microglial arms make direct contact with axons and dendrites,[25,26] implying that microglia may be carefully listening in on nerve cell conversations. To see if this is true, scientists devised experiments to test whether microglia reacted to what they "overheard." Indeed, when neuronal activity in the visual cortex was reduced (by maintaining the young mice in darkness), the microglia paid less attention to (made fewer contacts with) those neurons that normally would have received input about light signals, presumably because those neurons were talking less.[25,26] In contrast, increasing neuronal activity (by repetitive visual stimulation)

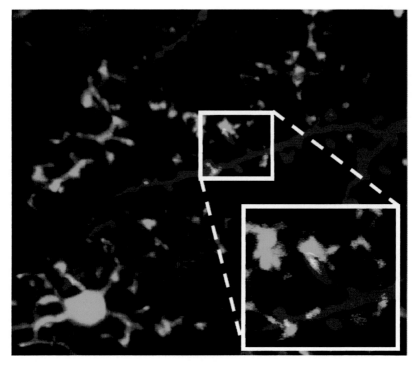

Figure 2. Microglia dynamically interact with synaptic elements in the healthy brain. Two-photon imaging in the olfactory bulb of adult mice shows processes of CX-3CR1-GFP-positive microglia connecting to tdTomato-labeled neurons. Reprinted with permission from Jenelle Wallace at Harvard University (Hong and Stevens, 201620).

resulted in increased contact by the microglial arms, which preferentially contacted and wrapped around neurons with high activity and energy use. Contact by microglial processes was associated with a subsequent decrease in spontaneous neuron firing, which may be a homeostatic response.

Due to their ability to listen to synapses and their role as macrophages (which are good at engulfing and eating things), many scientists wondered whether microglia might also play a role in synaptic pruning.

Developmental Synaptic Pruning

As the brain develops in the womb and during childhood and puberty, it needs to be gradually and carefully re-wired, with unneeded synapses removed or re-routed to more appropriate targets. This synaptic pruning is carried out, in part, by microglia.[22,27] Indeed, electron microscopy and high-resolution assays have found the remnants of synapses digesting within microglia in the mouse visual system, hippocampus, and other brain regions during the critical periods of synaptic pruning, the first few weeks of life in mice. In the visual system, as with all sensory systems, this pruning is dependent on neuron activity and sensory experience,[22,25] with microglia preferentially eliminating less-active synapses. But how do they know exactly which synapses to eat?

The nervous and immune systems share an array of molecules that have both specialized and analogous functions. Surprisingly, several proteins associated with innate (generalized) and adaptive (highly specialized) immunity are found in synapses, where they regulate circuit development and plasticity.[28-30] Among these substances are components of the classical complement cascade, which coat troublesome cells such as bacteria with "eat me" signals that attract macrophages that then engulf and digest them. A key molecule in this process is called C3. In the healthy developing mouse brain, C3 is widely produced and localizes to subsets of immature synapses.[31] There, it attracts microglia, the only nervous system cell type that has a C3 receptor, which then engulf the synapse, much as macrophages destroy bacteria outside the brain (Figure 3).[22] Mice without C3 and other proteins in this pathway have too many synapses and develop sustained defects in neuronal connectivity and brain wiring. Such excessive connectivity could result in increased excitability and seizures, as was demonstrated in mice that lack another protein in the complement pathway, C1q.[32]

There are likely other immune-related molecules (one is a sort of small, signaling cytokine or "chemokine" called Fractalkine[33]) that work in concert with the complement cascade to ensure that the right synapses are pruned at the right time. It is possible that different mechanisms regulate pruning in different contexts, e.g. across brain regions and stages of development. Aberrant pruning during developmental critical periods could contribute to neurodevelopmental disorders, such as autism and schizophrenia, as discussed on the following page. Indeed, emerging genetics identifies variants in complement protein C4 that increase the amount of complement in

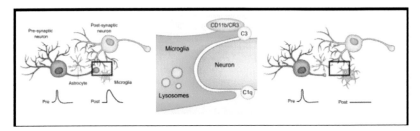

Figure 3: Synaptically coupled (i.e. communicating) neurons are under constant surveillance by glial cells, including microglia. If a neuronal synapse becomes "tagged" with complement protein C3, microglia recognize the tag with their C3 receptor (CR3/CD11b). This signal tells the microglia to engulf, or phagocytosis, and degrade the synapse. After microglial synaptic pruning, the eliminated synapse changes the way neurons communicate. Adapted from Lacagnina et al., 20173.

the brain and the risk of developing schizophrenia,[34] suggesting a model in which too-much-of-a-good-thing results in defective brain wiring.

Implications for Disease

As suggested above, synaptic pruning is a sensitive process; destroying too many or too few synapses will be detrimental. Factors in the environment, such as infectious disease, and within a person's own genome, such as mutation, may affect microglia's ability to find and destroy the appropriate synapses, leading, perhaps, to psychiatric conditions such as autism or schizophrenia, or neurodegenerative diseases such as Alzheimer's disease. Since they have complex and diverse functions in the brain, there are likely many ways in which microglia might contribute to disease risk and pathogenesis. Understanding when and where they become dysfunctional in these disorders will be critical to understanding how they influence relevant circuits and brain regions. Targeting the mechanisms that are dysregulated has the potential to arrest or reverse neurodevelopmental and neurodegenerative disorders where these cells play a role.

Early-Life Immune Activation

Microglia are immune cells, and thus respond to infection and inflammation. This may interfere with their normal duties (for instance, synaptic pruning), particularly if those infections happen during a critical time in brain development. Microglia develop slowly over normal embryonic and postnatal development; they start out as round, macrophage-like cells and gradually transform into the mature cell type illustrated in Figure 1. The functional implications of this shift in cell shape and structure are not fully understood, but many disorders are associated with strange-looking microglia. For instance, amoeboid (round) microglia are found in the post-mortem brains of autistic patients, even in later life, at a time when the cells should have long, thin processes, suggesting dysfunction in these cells.[35-37]

Studies with rats have shown that bacterial infection in newborn rats strongly activates the immune system, and that in young adulthood their microglia look round and dysfunctional, like those in the brains of patients with autism.[38,39] These newborn-infected rats also exhibit social deficits[40] and profound problems with learning and memory in adulthood—but only if they receive an injection of lipopolysaccharide (LPS), a key component in bacterial cells, around the time of learning.[39,41] This "second hit" apparently reactivates the immune system, which kicks the microglia into overdrive, overproducing a cytokine called IL-1β. This compound is vital for normal synaptic function and the formation of memories, but too much impairs memory.[39] So, the microglia of rats activated by infection as newborns act a bit like unruly teenagers weeks later, overreacting to the slightest provocation and causing problems.

Because microglia are long-lived cells (with slow turnover, about 28 percent per year in humans[42]) and can remain functionally activated, these insults early in life may persist into the future. Many additional studies with rodents have demonstrated that diverse inflammatory factors beyond infection, including stressors or environmental toxins, may similarly cause persistent changes in microglia that impact adult behavior.[43] Could such early-life insults—and their effects on microglia—result in serious problems much later in life?

Microglia in Neurodegenerative Disorders

Activated microglia and neuroinflammation are hallmarks of AD and other neurodegenerative diseases, including Parkinson's disease (PD), amyotrophic lateral sclerosis (ALS), and frontal temporal dementia.[44] These hallmarks were long considered to be symptoms rather than causes of disease, but new genetic studies indicate that they are indeed important, as many genes that increase risk of developing AD are enriched or specifically expressed in microglia.[45]

Microglia have complex roles that can both attenuate and exacerbate

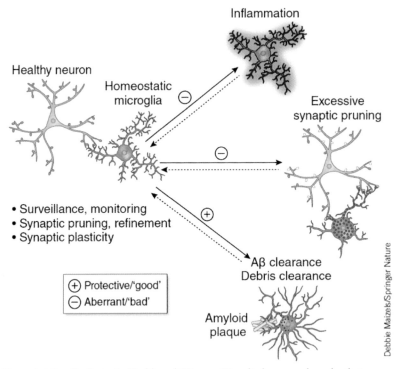

Figure 4. Microglia States in Health and Disease: Microglia have complex roles that are both beneficial and detrimental to disease pathogenesis including engulfing or degrading toxic proteins (i.e., amyloid plaques) and promoting neurotoxicity through excessive inflammatory cytokine release. Aberrations in microglia's normal homeostatic functions (Surveillance, synaptic pruning and plasticity) may also contribute to excessive synapse loss and cognitive dysfunction in AD and other diseases. Salter and Stevens 2016[46] with permission.

AD's pathogenesis. When the AD brain is cluttered with toxic amyloid plaques, microglia surround them, engulfing or degrading them and secreting inflammatory cytokines in the process (Figure 4).[46] Failure to clean up the dying cells, cellular debris, and toxic proteins like the amyloid plaques would contribute to inflammation and neurodegeneration. But overproduction of cytokines by microglia is also harmful. And excessive engulfment of synapses by microglia might contribute to cognitive impairment in AD.[20,47,48]

Synapse loss is in fact a hallmark of AD and many other neurodegenerative diseases, and can occur years before clinical symptoms—and fewer synapses in the AD's brain correlate with cognitive decline.[49,50] The mechanisms underlying synapse loss and dysfunction are poorly understood, although there are clues. Classical complement cascade proteins—the "eat me" signal involved in developmental pruning—are abundant in mouse models of AD in the hippocampus and vulnerable brain regions, binding to synapses before overt plaque deposition and signaling microglia to destroy those synapses. Similarly, recent evidence suggests that complement activation and microglia-mediated synaptic pruning contribute to neurodegeneration in mouse models of frontal temporal dementia,[51] glaucoma, and other diseases.[31,52]

These findings imply that the same pathway that prunes excess synapses in development is inappropriately activated in AD and may be a common mechanism underlying other neurodegenerative diseases. Thus, understanding the signals that trigger microglia to prune vulnerable circuits could provide important insights into these diseases and novel therapeutic targets. Given the diverse and complex activity of microglia in the healthy and diseased brain, there is a critical need for new biomarkers that relate specific microglial functional states to disease progression and pathobiology. Newly developed approaches to single-cell RNA sequencing and profiling of rodent and human microglia are likely to be fruitful here.

The Way Forward

We are just beginning to understand how microglia work in health and disease. But what we already know of their diverse roles in the healthy nervous system strongly suggests that some neurodevelopmental[53] and neurodegenerative disorders result in part from their dysfunction. Targeting these aberrant functions, thereby restoring homeostasis, may thus yield novel paradigms for therapies that were inconceivable within a neuron-centric view of the brain. But recent findings about the varying roles of microglia have come primarily from research with mice and rats, and it will be critical to understand which translate to humans. An investment in developing new models of disease, including human cell models,[46] is an essential next step toward clarifying whether the microglia-targeted therapeutic approaches emerging from rodent studies can, in fact, be used to treat human diseases.

12

Neuroimaging Advances for Depression

By Boadie W. Dunlop, M.D. and Helen S. Mayberg, M.D.

Boadie W. Dunlop, M.D., is an Associate Professor of Psychiatry and Behavioral Sciences and director of the Mood and Anxiety Disorders Program at Emory University. His primary research interest is in the application of biomarkers for use in personalized medicine for depression, posttraumatic stress disorder, and anxiety disorders. His other research interests include testing investigational medications for these disorders, and in the design and conduct of clinical trials. Dunlop also serves as the medical director of the Emory Healthcare Veterans Program and supervises a psychopharmacology specialty clinic as part of the psychiatry residency training program at the Emory University School of Medicine.

Helen S. Mayberg, M.D., is Professor of Psychiatry, Neurology and Radiology, and the Dorothy Fuqua Chair in psychiatric imaging and therapeutics at Emory University. Her research has characterized neural systems mediating major depression and its recovery, defined imaging-based illness subtypes to optimize treatment selection, and introduced the first use of deep brain stimulation for treatment resistant patients. She is a member of the National Academy of Medicine, the National Academy of Arts and Sciences, and the National Academy of Inventors, and has authored more than 200 publications, and participates in a wide variety of advisory and scientific activities across multiple fields in neuroscience.

Editor's Note: Depression is one of the world's most prevalent mental health problems, with as many as 350 million sufferers worldwide and close to 20 million sufferers in the US. While neuroimaging applications for identifying various types of depression have made enormous strides in recent years, no findings have been sufficiently replicated or considered significant enough to warrant application in clinical settings. Our authors are well equipped to tell us what the future may bring.

FOR THOSE WHO SUFFER FROM DEPRESSION, core symptoms of low mood, lack of motivation, and mental clouding are miserable and frustratingly difficult to overcome. For those treating it, the complexity arising from varied combinations of these and other symptoms[1] typically leads to the default application of treatment algorithms based on "average" patient outcomes.[2] Patients and clinicians alike would benefit tremendously from methods using specific clinical or biological features to personalize treatment; i.e., to select the approach most likely to help. This goal is the focus of much of today's depression research.

Depression is best thought of as a syndrome, which takes diverse forms in different people. All sufferers share the symptoms of excessive sadness or lack of pleasurable experience, but the associated characteristics can vary widely. Some have increased appetite and sleep, while others experience the opposite. Problems with concentrating or making choices may plague some, whereas others get bound up in guilty ruminations about past actions. Of greatest concern is that some patients—not necessarily those with severe symptoms and impairment—develop thoughts of taking their lives. All this variability suggests that there should be ways to classify depression into subtypes based on specific biological disturbances.

Clinical Classifications of Depression

The ultimate goal of classifying depression into subtypes is to enable selection of treatments that will hasten the resolution of the illness and, ideally, prevent future episodes. Psychiatrists recognize the importance of classification, and have expended tremendous time and energy on parsing patients' clinical features, including symptoms, age of onset, frequency of recurrent episodes, and the role of preceding events.[3-5] The most important diagnostic factor to emerge has been the identification of hypomanic or manic episodes when the patient is not depressed, which serves to distinguish between bipolar disorder and major depressive disorder.[6]

Prominent among other clinical features used for subtyping major depressive disorder are the presence of psychotic features, peri- or post-partum onset, and seasonal variation. Many other bases for clinical categorization have been proposed over the past century, including endogenous versus reactive or neurotic depression, anxious versus non-anxious features, and melancholic (i.e., unrelentingly low mood associated with insomnia and reduced appetite) versus atypical features (i.e., having brief periods when the mood substantially lifts associated with increased appetite and sleep)

Of these, perhaps the most clinically relevant for treatment selection are: 1) the identification of psychotic symptoms that require the addition of an antipsychotic to an antidepressant, or electroconvulsive therapy, and 2) the prominence of atypical features, which respond better to serotonin reuptake inhibitors (SSRIs) or monoamine oxidase inhibitors than to tricyclic antidepressants.[7] Although DSM-5 (the standard classification of mental disorders used by mental health professionals in the US) added many additional "specifiers" to more fully characterize major depressive disorder, their utility for treatment selection has not been established. Efforts to further classify depression continue, using increasingly complex models to group and weight specific symptoms and demographic features that can predict treatment outcomes.[8-11] Unfortunately, these approaches to clinical subtyping have had limited value for personalizing treatment approaches due to inadequate or unsuccessful replication studies.

Similarly, biological studies of major depression have not yet met with great success in subtyping the disorder. Evaluations of neurochemistry, genetics, electroencephalography, neuroendocrinology, inflammation, and metabolomics (the study of molecules in the blood derived from cellular processes) have been investigated, with inconsistent results. High levels of inflammatory markers are of particular interest as a potential etiology of "treatment-resistant depression," i.e., an episode that has not responded to established depression treatments. Some preliminary analyses suggest that a biological marker of inflammation can help indicate the most appropriate treatment,[12-14] though further validation is still required. Many researchers believe that advances in understanding the biology of depression will require combinations of biomarkers. For example, it is possible that genetic or chemical markers derived from blood samples may help classify depression most effectively in conjunction with brain-focused methods of research.

Neuroimaging Toolkit

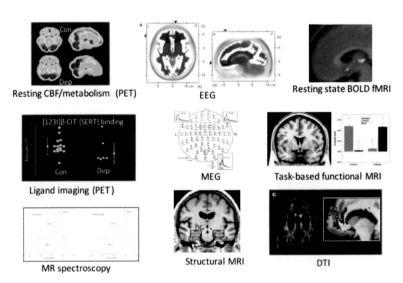

Figure 1. Above are seven of the neuroimaging forms related to mood disorders: Positron emission tomography (PET): electroencephalography (EEG); blood oxygen level-dependent (DTI: Diffusion tensor imaging; Cerebral blood flow (CBF); Functional magnetic resonance imaging (fMRI); and magnetoencephalography (MEG).

Neuroimaging Applications to Depression

Neuroimaging may provide the greatest hope for identifying depression subtypes.[15] There are many forms of neuroimaging that have relevance to mood disorders and these have become increasingly more sophisticated over time (see Figure 1).

Initial studies examined the structure of the brain, particularly the size or volume of specific brain regions, using static images obtained via computerized tomography (CT) or magnetic resonance imaging (MRI). These studies identified the frontal cortex and hippocampus as potentially relevant brain regions in the pathophysiology of depression, hypotheses that were supported by post-mortem studies.

Subsequent research looked at regional blood flow or energy metabolism in the brain using functional MRI (fMRI) or positron emission tomography (PET). These studies capture the activity of the brain either in the "resting state" (i.e., when patients are not focusing on any particular thought or stimulus) or when the brain is actively responding to a task that induces an emotional or cognitive response. More recently, researchers have applied machine learning methods to fMRI data to identify brain networks and connectivity.

Other types of neuroimaging for mood disorders include diffusion tensor imaging (DTI) to assess the integrity of the white matter tracts that connect regions of the brain, magnetic resonance spectroscopy to assess differences in chemical composition across these regions, and ligand-binding studies to measure the density of receptors or monoamine-transporters.[16] There are clearly many approaches to imaging the brain, and many interesting models of brain function and dysfunction in depression have been developed. The challenge facing researchers today is how to reliably establish and clinically apply these various conceptualizations.

Efforts to define biomarkers that can reliably distinguish between patients with depression and healthy controls have included neuroimaging.[17,18] The earliest neuroimaging studies of major depression identified several differences between non–depressed and depressed individuals. Many studies using resting state or task-based PET or fMRI found depressed patients to have reduced metabolism in the dorsolateral prefrontal cortex, along with increased activity in regions considered part of the limbic system, such as the insula, amygdala and subcallosal cingulate cortex.[19]

A recent meta-analysis of resting state functional connectivity studies identified reduced connectivity in frontoparietal control systems in depressed versus healthy control subjects, though without sufficient accuracy for a diagnostic test.[20] Meta-analyses of structural MRI studies have shown, on average, smaller volumes of the hippocampus or its subregions in depressed patients,[21,22] along with cortical thinning in several brain regions relevant to mood processing.[23] In some studies, smaller hippocampal volume was associated with poorer outcomes to antidepressant medication treatment, though other studies have not replicated this volumetric finding or their treatment-predictive effects.[24,25]

In sum, none of these approaches have achieved the level of accuracy or reproducibility that would define at an individual level the "neural signature" of major depressive disorder."[2,27] The core of the problem is the clinical and biological heterogeneity within the diagnosis, along with high rates of comorbidity with other psychiatric disorders.[28]

Neuroimaging–Based Depression Subtypes

Despite its failure to distinguish the healthy from the depressed, neuroimaging may have value in defining subtypes of depression. A crucially important application would be distinguishing between bipolar and major depressive disorder among patients presenting with a first episode of major depressive disorder, because misdiagnosis is common and the treatments for these two disorders are quite different.[29] Neural markers purported to accurately classify these patients have used structural imaging,[30,31] resting state functional connectivity,[32] emotion–eliciting tasks,[33] and white matter integrity.[34] Although these methods have identified differences at the group level, their inability to clearly distinguish individuals with bipolar disorder from those with major depressive disorder may indicate that these illnesses represent points on a spectrum, rather than distinct biological entities.[35]

Given the difficulty in generating markers that can reliably sort major depression and bipolar depression, it may seem unlikely that neuroimaging will achieve clinically relevant subtyping within major depression alone. However, there are many interesting neuroscience theories with potential relevance here. Several neuronal circuits (i.e., interconnected brain regions that work together during specific mental functions) have been proposed that, when dysfunctional, may result in the symptoms of depression and anxiety.[36] Researchers have now begun rising to the challenge of defining which neuroimaging markers have utility for treatment selection at the individual level. This leap forward will require imaging datasets subjected to an iterative process of identification, replication, and testing for clinical utility. The most fruitful path for applying neuroimaging to achieving this goal remains to be determined.

Various Approaches

Researchers attempting to subtype major depressive disorder with neuroimaging have taken several approaches. One has been to examine traditional clinical subtypes, such as melancholic, psychotic, or atypical features, for their neuroimaging correlates.[37-40] Most such studies have found no or limited support for underlying neural signatures for these categories.[41] Small studies have suggested that melancholic depression is specifically associated with abnormal function of the default mode network (i.e., a brain network that is active when a person is not directing their attention to external events)[42] or reduced effective connectivity between the insula and attentional networks.[43] Overall, although they have interest, approaches that seek links between the brain and classic symptom features do not seem to offer promise for moving toward the goal of treatment selection.

A second approach has been to examine core features of depression using task-based studies, which analyze patterns of reactivity in the brain as the patient focuses on a specific cognitive or emotional stimulus. For example, anhedonia (the loss of motivation or pleasure in activities) is common in depression, although not all patients experience this symptom. Young people who demonstrate reduced neural responses to rewarding stimuli have been accurately identified as being at risk for developing depression.[44,45]

Other task-based assessments that presented emotionally salient words or faces to assess the reactivity of limbic (especially amygdala, and prefrontal) emotion-regulation regions have had discrepant results. A recent meta-analysis of 57 studies that employed emotion- or cognitive-processing tasks failed to find consistent differences between depressed and healthy control patients in brain activation patterns.[46] Because few of these studies have been large enough to identify differential patterns among depressed patients, the potential for subtyping with these methods is unclear.[47,48]

A third, more ambitious approach to subtyping has worked from the bottom-up, without labeling the subjects *a priori* but using unsupervised machine learning methods to find inherent patterns of connectivity or reactivity in neuroimaging data. The patterns identified from these analyses can then be used to cluster patients, creating potentially novel depression subtypes. Such analyses require large datasets to ensure that these patterns don't simply characterize the participants in the study, but can be applied to depressed patients more generally.[43,49]

Recently, the largest study of this kind, published in *Nature Medicine*, used resting state functional connectivity data in 220 patients with treatment-resistant depression to define biological subtypes ("biotypes") of major depression.[50] This analysis identified four forms of dysfunctional connectivity between fronto-striatal and limbic networks, which demonstrated high classification accuracy and were differentially associated with clinical symptoms of anhedonia and anxiety. They were replicated in an independent sample. The clinical utility of these biotypes remains uncertain because treatment outcome data for most patients were not available, though in a subset of the patients the biotypes were retrospectively able to identify those who were likely to respond to treatment with a form of transcranial magnetic stimulation.[50] This approach of building a new categorical system from the bottom up, uninfluenced by prior conceptualizations of depression, carries the appeal of carving nature at its joints, but will likely require decades to be sufficiently validated to use in treatment selection.

In contrast to the grand scheme of characterizing *de novo* the pathophysiology of depression for the purposes of subtyping, and hoping that treatments will follow those subtypes, it may be more useful to take a direct approach: i.e., to explicitly apply the treatment outcomes of patients to the neuroimaging data. Through this fourth approach, the treatment outcomes (e.g., remission, non-response) are used to identify signatures in the pre-treatment neuroimaging data that differentiate between the outcomes. These neuroimaging signatures can then be prospectively tested by seeing how well they predict outcomes in new patients receiving the treatments. Thus, this approach takes the stance that the maximal utility from brain imaging in depression will emerge from its ability to help clinicians choose between possible treatments, not from developing depression classification schemes isolated from treatment effects. Identifying neuroimaging patterns that can predict with reasonable accuracy the probability that an individual will benefit from a specific treatment would be a substantial boon to the clinical care of depressed patients.

Looking Forward

Several studies using resting-state or task-based fMRI have reported markers predictive of response or non-response to antidepressant medications[51-53] or to psychotherapy.[54-56] A limitation of these single-intervention studies is that they do not provide information about whether an alternative treatment would likely be more or less effective than the evaluated one. This makes it impossible to determine whether the neuroimaging signal in question indicates response regardless of treatment, or is specific to the intervention in the study.[57]

This treatment outcome-based approach would be most valuable if it could inform the prediction of good and poor outcomes to two or more treatments believed to work via differing mechanisms (e.g., psychotherapy versus medication versus TMS).

Using this approach, we recently published the results of two randomized studies that aimed to identify neuroimaging patterns that could differentially predict outcomes to treatment with an antidepressant medication or cognitive behavior therapy (CBT). Building off clinical observations that patients who responded poorly to one of these interventions often did well with the other, we hypothesized that neuroimaging patterns that predicted remission with one treatment would also predict poor response with the other.

The first study, using fluorodeoxyglucose-PET, found that resting state metabolism of the right anterior insula could distinguish remitters from non-responders to treatment with the antidepressant escitalopram or to CBT.[58,59] More recently, a subsequent study of 122 depressed patients used resting state fMRI to identify functional connectivity patterns between the subcallosal cingulate cortex and three other brain regions that could distinguish between remitters and non-responders to antidepressant medication (escitalopram or duloxetine) and to CBT.[60] These results support the concept that patients whose depression symptoms are similarly severe can have distinctly different patterns of brain activity, and that those patterns may be used for treatment selection. With this approach, it may be possible to predict failure to all standard first line treatments for depression, avoiding months of ineffective therapies and allowing earlier application of interventions usually reserved for treatment-resistant depression, such as TMS, electroconvulsive therapy, ketamine, or medication polypharmacy.[61]

In sum, despite the excitement around neuroimaging methods, it is important to recognize that to date all the identified subtypes have been based on retrospective analyses. The great majority of patients already benefit from one or another of existing depression treatments. But approaches that characterize brain states responsive to specific interventions offer the possibility of advancing further, past the current trial-and-error, algorithm-based application of psychotherapy, medication, and brain stimulation to a personalized psychiatry that can choose the treatment most likely to benefit the individual patient.

BOOK REVIEWS

13

Eric Kandel's *Reductionism in Art and Brain Science* – Bridging the Two Cultures

Review by Ed Bilsky, Ph.D.

Ed Bilsky, Ph.D., is the provost and chief academic officer at Pacific Northwest University. He is a neuropharmacologist by training, having spent the last 30 years studying the neurobiology of opioids and opioid receptors, and their connections to pain, substance use disorders, and other complex behaviors. He is a National Institutes of Health funded professor of biomedical sciences, author of more than 80 peer-reviewed publications, and frequent speaker. He is passionate about higher education, student centered research and scholarship, and engagement with K-12 school systems and the public on important health issues involving the nervous system. He is a member of the Dana Alliance for Brain Initiatives and a committee member for the public education and communication committee of the Society for Neuroscience.

Editor's Note: Reductive art is a term to describe an artistic style or an aesthetic, rather than an art movement. It is stripping down as a new way of seeing. Movements and other terms that are sometimes associated with reductive art include abstract art, minimalism, ABC art, anti-illusionism, cool art, and rejective art. Eric Kandel's fifth book focuses on reductionism as the principle guiding an ongoing dialogue between the worlds of science and art.

WHILE ATTENDING THE SOCIETY FOR Neuroscience annual meeting this past fall, I came across advertisements for the release of *Reductionism in Art and Brain Science – Bridging the Two Cultures* and the opportunity to have it signed by the author, Eric Kandel. Arriving in Publisher's row, I stood in a long line, a testament to the reverence the neuroscience community has for the Nobel laureate. By the time I got to the booth, the publisher was down to its last two copies, leaving many disappointed fans behind me.

Months passed before I carved out the time to read it. Not surprisingly, I was soon engrossed; Kandel is known for clearly communicating complex concepts, whether in a definitive neuroscience textbook (*Principles of Neural Sciences*) or wide-ranging forays into the interface of science, art, and culture (*Age of Insight*). *Reductionism In Art and Brain Science*, while more compact, is no less ambitious. Helping draw the reader in is the quality of the printing. The book has a solid feel and the pages are on high quality paper appropriate for the stunning images Kandel selects to help tell his story.

Coming into his narrative, I had a solid grasp of the neural principles of the visual system, learning and memory, and cognitive processing. I was, however, a near-neophyte when it came to art and art history. The author elaborates (or should I say reduces?) the relationship between how the human nervous system processes and perceives the visual world, while also highlighting the dramatic changes taking place in the world of art that began in the 19th century and continues today.

Kandel believes that photography was a disruptive innovation that threatened the realists who had developed extraordinary techniques to evoke in intricate detail a three-dimensional landscape or portrait on a two-dimensional canvas. The artistic response evolved rapidly, even within the lifetime of artists who adapted their styles to meet the challenge. The choice of art featured in each chapter complements Kandel's clarity of writing and crystalizes the points he is attempting to make.

Joseph Mallord William Turner's seascapes from 1803 and 1842 exemplify the origins of this perspective, illustrating some of the first attempts to replace detail with elements of abstraction to create even more evocative works of art. Another great set of illustrations features works of Austrian composer and music theorist Arnold Schoenberg as his approach to portraits evolves in the span of a single year.

One of the most satisfying parts of the book is the discussion of bottom-up and top-down processing of information by the nervous system. The evolution of visual systems that allow organisms as simple as insects and as complex as human beings to interact effectively with a dynamic, fast changing three-dimensional world in real-time is simply astonishing. The near "hard wired in" systems that enable scanning an open field and tree line for predators and prey, or recognizing faces and facial expressions in one's tribe or village, are crucial for survival. But as cortical regions developed and became more complex, the brain was also evolving the capacity to be curious, solve puzzles, and fill in the blanks when presented with incomplete information.

When confronting an abstract piece of art, the viewer's nervous system is challenged to function in a novel way—to scan and assimilate visual information that does not add up to the kind of representational image it evolved to interpret, and to understand this information through one's own lens and unique framework. This can certainly be frustrating— seemingly impossible, even—with some pieces of art, but as in many other situations, the nervous system can learn with repeated exposure and benefit from the mentorship of one who is more experienced. Kandel is extremely adept at explaining what seem unlikely connections—between discoveries in our understanding of the nervous system and how we perceive and react to our surroundings, and how artists were adapting and evolving to engage audiences in new and provocative ways.

The utility of reductionist approaches to both neural sciences and to the creation of visual art is well documented as the reader works through each chapter. Kandel draws on his own research using simpler systems in which to study fundamental processes of learning and memory, allowing for eventual reconstruction into complex systems of human cognition in both health and disease. On the artistic side, the creation of abstraction—breaking the visual world into its basic elements of line, light, form, and color—is discussed and fully illustrated. The cover art by Mark Rothko is just one example of this component.

That Kandel's book made me a more open-minded and curious viewer of art became apparent shortly afterwards, when I visited the Smithsonian American Art Museum in Washington, DC. with my teenage children. I sought out specific pieces in the collection and spent more time with those pieces than I would have otherwise. The adventure was enhanced by listening to my kids' comments and observing their level of engagement and patience in each section of the museum. I did my best to explain some of the historical context and interpretation I had gleaned from the book and apply them to the exhibits.

A good book is memorable while a great book makes you think differently about a topic and perhaps change a behavior or apply what you have learned to a new challenge. I found myself excited to engage with both fellow neuroscientists and artists within the university faculty—thanks, in part, to my newfound enthusiasm for the topic and for sharing the experience of reading the book.

But another question has developed over the past few months, as I have gone back to reread sections of the book and view and revisit more of the artwork. It crystalized in a discussion with scientific colleagues. What value do members of the lay public perceive in a reductionist's understanding of learning and memory when a parent has dementia? How is that piece of abstract art going to help me afford college for my child or enhance his or her education?

One response is to offer a foothold for understanding science and the arts to a broader swath of the public. As people experience new things and gain confident familiarity with the topics they represent, this enhances rather than stifles their curiosity. Inviting people into these fascinating fields and giving them a sense of how they bleed together to spark innovation and address complex problems is another of the book's significant contributions.

We desperately need more authors like Kandel—thinkers who can master a discipline, move comfortably between the arts, sciences, and cultures, and communicate effectively with the public. We need scientists, artists, and scholars willing to engage with people without college degrees, children in diverse school systems, and policy makers who all too often hold too narrow a view of the world. Kandel has done a masterful job in "bridging the two cultures" of art and science through a reductionist approach. He also opens the door for us to build back up to the level of the individual, the family, and, the community. This may in fact be a viable approach in which to form new bridges across cultures that are isolated and deeply divided. These are doors we need to enter, and through his writing, Kandel demonstrates how science and art can be catalysts for positive change.

14

Joseph J. Fins' *Rights Come to Mind: Brain Injury, Ethics and the Struggle for Consciousness*

By Arthur L Caplan, Ph.D.

Arthur L. Caplan, Ph.D., is the Mitty Professor and founding head of the Division of Medical Ethics at NYU School of Medicine. Previously, he was the Caplan Professor of Bioethics at the University of Pennsylvania Perelman School of Medicine, where he created the Center for Bioethics and the Department of Medical Ethics. Caplan has also taught at the University of Minnesota, the University of Pittsburgh, and Columbia University. The author or editor of 35 books and over 725 papers in peer reviewed journals, Kaplan's most recent book is *The Ethics of Sport*, (Oxford University Press, 2016). He has served as the chair of the National Cancer Institute Biobanking Ethics Working Group, chair of the advisory committee to the United Nations on Human Cloning; and chair of the advisory committee to the Department of Health and Human Services on Blood Safety and Availability. Caplan was a *USA Today* 2001 "Person of the Year" and was described as one of the ten most influential people in science by *Discover* magazine in 2008. He received his Ph.D. from Columbia University.

Editor's Note: The book reflects Fins' role as co-director of the Consortium for the Advanced Study of Brain Injury at Weill Cornell Medicine and the Rockefeller University and his struggle to answer the kinds of questions that stand to shape how society treats people with brain injuries. What is the capacity of brains to recover? What are the mechanisms of that recovery? How do we know that our assessments are accurately describing what's going on in a patient's mind? And what does society morally owe these patients and families?

I NEED TO MAKE TWO THINGS CLEAR at the start of this review. One is about conflict of interest: Joe Fins is a good friend of mine. So why would I review the book of a good friend? That leads to the second thing: *Rights Come to Mind* is a wonderful book; perhaps the best book ever to emerge from the young field of neuroethics. By any criteria, the book is an inspiring exemplar of how to integrate ethics and medicine. When a book is so manifestly outstanding, the reviewer's conflicts of interest wither away.

What makes the book so good? To start, Fins writes clearly and in an orderly, organized manner. From chapter to chapter he tells you what he is going to do, does it, and then recapitulates what he did and why.

Fins studied with giants like Fred Plum, a neurologist whose pioneering research advanced the understanding and care of comatose patients. (Plum coined the term "persistent vegetative state.") As a result of his training, Fins knows his way around the injured brain and is a thoughtful guide to those who do not.

The book starts with a thorough, insightful examination of the history of medicine's and science's understanding of brain function and brain injury. It progresses to a series of cases, some in depth, others in snippets, all engaging and illuminating as they reveal the struggle those caring for their loved ones have had and continue to have in trying to do what is best for the severely brain injured.

Fins is the motive force behind the notion that the permanent vegetative state includes less irreversible and impairing variants, which he terms 'minimally conscious.' He builds on recent advances in imaging, in the understanding of brain injury etiology and effects, and in deep brain stimulation to make a case that not all non-responsive comatose patients are alike. He argues that 'prognostic pessimism' and 'therapeutic nihilism' in the face of non-responsiveness need to be replaced with more patience, better care, and more intense efforts to determine if an individual capable of thought and perception still resides inside a damaged brain.

Which leads to the ethical heart of this important book: Fins wants to return rights to brain injured individuals once thought to be permanently vegetative but who may well be minimally conscious. He argues that recovery from a terrible brain injury may need both better rehabilitative settings and more time. He proposes that drugs or neuromodulation by direct brain stimulus may lead to either brain cell regeneration or the activation of other pathways that might permit more integrated brain activity and patient communication. In other words, he wants us to view the minimally conscious as a new category of patient who merit both research funding and resources for clinical care.

His convincing case leaves me worried.

While being minimally conscious may allow more cognition of some sort than being irreversibly comatose and unconscious, it is far from clear that this is a better state to be in. Finding out that one is at best dimly aware of one's surroundings and cannot communicate to anyone, and that one's body is unable to perform any but autonomic tasks might well fill a person with dread. The understandable sense of horrific dread many people feel in thinking about finding themselves or a loved one possibly facing end-stage ALS or locked-in syndrome, in which a stroke or aneurysm leaves a person fully aware but unable move or communicate due to complete paralysis, may not lead them to demand better clinical care in a specialized setting but rather to demand that care be ended.

What those who complete advanced directives will say they want done—given the current state of knowledge about the minimally conscious state—is not necessarily what Fins deems appropriate. Nor is it clear from his argument what clinicians ought to say to a patient's loved ones, or even to a patient who they suspect might be minimally conscious, in terms of the continuation of care.

And while research to reverse the effects of brain injury is progressing, and new tools to control the problem have emerged, in fact we neither understand the brain well enough to truly know what we are doing in this area nor have any idea how long effects induced by research interventions will last. Calls for research are likely to be heard by desperate families as opportunities for novel therapies. Before venturing further down the research path, it is important to call more loudly than Fins does for sound clinical trials, competent investigators, national and international registries, diagnostic homogeneity, conflict of interest management, and rigorous Institutional Review Board examination.

And then there is the question of the right to health care for the minimally conscious in a nation like the United States, where the *fully* conscious cannot be assured of life-saving or disability preventing care. Justice may forbid discrimination against those with even   severely incapacitating neurological conditions. But if there is no publicly funded long-term care insurance and no budget for home health assistance, and co-pays for rehabilitation services or novel treatments are prohibitive, then it is not likely that affording the minimally conscious the same rights as others will do them much good.

Like any provocative book, Fins' work offers plenty to argue about. What ought to be the consequence of acknowledging that thousands, maybe tens of thousands of people around the world may be in a minimally conscious state? My hunch is that it will take decades to figure out exactly who they are and how best to treat them. Fins wants a faster timetable. I may well be wrong, but without this thoughtful, compassionate, and principled book, I would never have realized my obligation to worry about who is right.

15

Alan Alda's *If I Understood You, Would I Have This Look on My Face?: My Adventures in the Art and Science of Relating and Communicating*

By Eric Chudler, Ph.D.

Eric H. Chudler, Ph.D., is a neuroscientist at the University of Washington and the executive director of the Center for Sensorimotor Neural Engineering in Seattle. In addition to conducting research related to how the brain processes information from the senses, he has worked with fellow scientists and teachers to create materials to help the public understand how the brain works. Chudler has conducted workshops and given presentations to a variety of audiences including precollege students, university students, teachers, judges, and Tibetan Buddhist monks and nuns. He has also published several books for general audiences, including *Brain Bytes: Quick Answers to Quirky Questions About the Brain* (W.W. Norton & Co., 2017) and *The Little Book of Neuroscience Haiku* (W.W. Norton & Co., 2013). Chudler graduated from the University of California, Los Angeles (UCLA) in 1980 with a B.S. degree in Psychobiology and attended the University of Washington, where he received an M.S. degree and a Ph.D. degree from the Department of Psychology. His post-doctoral work was performed at the National Institutes of Health.

Editor's Note: A primary function of my role is asking top neuroscientists to write about the latest developments in their specialty areas for lay readers. If they agree to the assignment, I encourage them to use—whenever possible—conversational language, anecdotes, storytelling, and their own voice in communicating what are often complex and hard-to-explain topics. Another option might be to suggest they read Alan Alda's new book before they begin.

Most people know Alan Alda as an actor on TV (*M*A*S*H*, *The West Wing*, *The Blacklist*), in film (*Crimes and Misdemeanors, The Aviator*) or on stage (*Glengarry Glen Ross, Love Letters*). Few may realize that Alda has also championed efforts to help scientists improve the way they communicate their work. Alda started this quest in 1993 as the host of the PBS television series Scientific American Frontiers and he continues this work at the Alan Alda Center for Communication Science at Stony Brook University. In his new book, *If I Understood You, Would I Have This Look on My Face?*, Alda describes his efforts to provide scientists and health professionals with tools to communicate clearly with the lay public.

As you might expect from a book about effective communication, *If I Understood You, Would I Have This Look on My Face?* is written in a way that is easy to understand. With humor and a clear, concise, and never stilted writing style, Alda takes readers on his journey to help experts convey neuroscience and other complex scientific topics to a variety of audiences.

Alda describes how his extensive experiences as an actor have given him skills that help him communicate with others. He explains how improvisational games and exercises can teach people to better understand themselves and their audiences. One such game, "mirroring," starts with one person moving and a second copying the movements. After a while, the movements of the two people become synchronized. Another exercise involves tossing an imaginary ball back and forth. These and other fun activities are designed to help people relate to one another. And that's a recurring theme throughout the book: to become a better communicator you must understand what other people are thinking. You must put yourself in the shoes of your audience and develop empathy.

The value of empathy for effective communication is underscored when Alda explains research he and others have conducted. He acknowledges that much of his own is anecdotal in nature, making it impossible to draw definitive conclusions about its significance. He admits that he does not know if his short workshops and talks have a long-lasting positive impact on participants' ability to communicate. But in summarizing the work of other researchers, some published in the scientific literature, Alda describes a growing body of evidence that confirms the value of empathy and effective communication. For example, sports teams that had better communication skills and more empathy were more successful; doctors who were rated as more empathic had patients who showed more gains in their health.

As a scientist, I would have liked the book to include citations to these studies, which would allow me to read the original papers and evaluate the research for myself. But then, the book is intended for a general audience, and perhaps Alda thought such a list would distract readers. And he should know; he and others have taught techniques to help thousands of people in workshops and events around the world to become better communicators. His work and that of trained researchers provide a rich source of theories and hypotheses about the keys to effective communication. Future experiments can test these ideas and help refine strategies to this end.

A motif that recurs throughout the research is that attentive listening and the ability to know whether an audience is engaged are essential for successful communication. To gauge your own level of empathy, Alda suggests the "Reading the Mind in the Eyes" test developed by Simon Baron-Cohen, in which people are shown photos of others in different emotional states. The twist is that test-takers are shown only the eyes of the other people, and must decide what emotion is being felt by each person from this alone.

When Alda took the test, he correctly identified the emotions in 33 of the 36 photos. I took it and scored the emotions correctly in 32 of 36. Interestingly, Alda improved his score by meditating and focusing on his breathing. This reinforced his belief that empathy can improve with practice and with attention to one's own feelings and thoughts. (You can find several versions of the Reading the Mind in the Eyes test online, to experiment on yourself.)

Alda states that he dislikes "tips" because they are weak unless they come with experience or story, but he provides many strategies useful for anyone who needs to express his or her point of view. Storytelling is one of the most powerful. Stories, especially those with emotional content, are particularly effective in creating long-lasting memories in listeners. In his book, Alda practices what he preaches by telling stories he has gathered over his years as an actor, host, and speaker. He also warns that jargon can alienate listeners. Special phrases and abbreviations have their time and place— when everyone in a group understands their meaning. But it is confusing and can waste valuable time when a speaker uses language that listeners cannot understand.

Each of us needs the ability to convey ideas well, whether to co-workers, family members, the media, students, or juries. Although Alda focuses on improving the communication skills of scientists and physicians, his book, *If I Understood You, Would I Have This Look on My Face?* will be valuable to teachers, students, lawyers, accountants, bus drivers—to those in just about all occupations. Even spouses, parents, and children who need help talking to each other can benefit from Alda's advice. We all need to communicate clearly.

16

Matthieu Ricard and Wolf Singer's *Beyond the Self: Conversations between Buddhism and Neuroscience*

Review by Paul J. Zak, Ph.D.

Paul J. Zak, Ph.D., is a scientist, author, and public speaker. His book *The Moral Molecule: The Source of Love and Prosperity* was published in 2012 and was a finalist for the Wellcome Trust Book Prize. He is the founding director of the Center for Neuroeconomics Studies and Professor of Economics, Psychology and Management at Claremont Graduate University. Zak also serves as Professor of Neurology at Loma Linda University Medical Center. He has degrees in mathematics and economics from San Diego State University, a Ph.D. in economics from University of Pennsylvania, and post-doctoral training in neuroimaging from Harvard University.

Editor's Note: Buddhism shares with science the task of examining the mind empirically. But Buddhism has pursued, for two millennia, direct investigation of the mind through penetrating introspection. Neuroscience, on the other hand, relies on third-person knowledge in the form of scientific observation. In the book that is the subject of this review, two friends, one a Buddhist monk trained as a molecular biologist, and the other, a distinguished neuroscientist, offer their perspectives on the mind, the self, consciousness, the unconscious, free will, epistemology, meditation, and neuroplasticity.

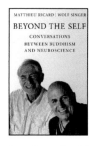

IMPULSIVELY, I RUBBED MATTHIEU RICARD'S head on the beach in Rio De Janeiro.

This is not as weird as it sounds. He is incredibly joyful, perhaps equally as much as his teacher Tenzin Gyatso, the 14th Dalai Lama. We were at TED Global and because I knew of Ricard's work, I had been trying to meet him, but he was constantly mobbed. When I spotted him walking alone, I said, "Matthieu, I have been looking for you!" He replied, "Paul, I have been looking for you." We hugged and talked science, and then the speaker concierge from TED tried to pull him away to do his "hair and makeup" preparation before his talk. Rubbing his bald head, I refused to let him get away from me that fast.

Having met Ricard and knowing Wolf Singer's research, I was delighted to learn of their book three years later. Ricard earned a Ph.D. in molecular biology and is a Buddhist monk who lives in a monastery in Nepal. He has participated in, and co-authored, neuroscientific studies of meditation, and *Beyond the Self* reflects his scientific training while drawing on his extensive study of Buddhist traditions. Singer is a physician and neurophysiologist who, until his recent retirement, led the Max Planck Institute for Brain Research in Frankfurt, Germany. Singer's research has focused on the neural basis for awareness and cognition.

Their book is an edited set of conversations Ricard and Singer had over the course of eight years at meetings around the world. This format offers an insight into how these two scholars have, with humor and skepticism, sought to find common ground between the third-person attempt of Western science to understand mental processes objectively, and the first-person introspective approach used over the last two millennia by Tibetan Buddhists. The reader is allowed to eavesdrop as the authors discuss fundamental issues in neuroscience—the nature of consciousness, how emotions are processed in the brain, and how practice changes mental processes—in an engaging and free-flowing way.

Ricard offers the effects of meditation on mental processes as a window into how the brain works, while Singer states the scientific consensus on the neural mechanisms producing these effects. Careful readers will gain much from their back-and-forth. Yet, because Singer is speaking to Ricard as one trained scientist to another, the most instructive points are at times glossed over and the authors use scientific jargon a more traditional book would expansively explain. Counterbalancing this is some repetition in explicating the neuroscience as the conversations start, stop, and start again; the redux aids comprehension as different aspects of key points come into the discussion. At its best, this format allows the reader to acquire knowledge about how the brain works in a natural way, without the dreariness of a textbook.

Of the two dialogists, Ricard is more fun, irreverent even, weaving captivating narratives that illustrate his 35 years of Buddhist practice. One particularly beautiful reflection that captures Buddhist tradition is, "We may also use attention to cultivate compassion. If the mind is constantly distracted even though it looks as if one is meditating, then the mind is powerlessly carried away all over the world like a balloon in the wind." This illustrates the yin and yang of the book: The scientist in me wanted the authors to identify the mechanism through which distraction inhibits compassion, but as an appreciative reader I simply reveled in the poetry of abstraction and metaphor.

Ricard understands the scientific studies of meditation and discusses them in detail, but ultimately he is more interested in its practice. The tension between understanding what meditation does in the brain and actually meditating is woven through the book and keeps the conversation lively. It also leaves the reader, and indeed, Singer, at a bit of a loss because he, and we, have presumably not meditated for thousands of hours and must take Ricard's account of his experience as genuine and generalizable.

While the book is focused on meditation, there is a significant amount of foundational neuroscience in it. For example, when the authors describe an experiment showing that experienced meditators can inhibit the startle reflex, they note that this reflex is generated in the brainstem and conventionally thought to be "autonomic" or outside of conscious control. The authors also discuss a study showing that experienced meditators can perform cognitively difficult tasks with less neural effort and for longer periods of time than can the untrained. This provides an insight into the extraordinary plasticity of the brain.

Both authors are humble in their dialog, consistent with the enthusiastic embrace of science by Buddhists and the centrality of compassion in their religion. Singer is the surrogate for the reader, probing Ricard with questions both personal and scientific. While he pushes Ricard on some topics, ultimately the third-person scientific approach cannot challenge Ricard's first-person experience as a monk because science is ultimately mechanistic while one's own experience is wholistic. Even though Ricard discusses scientific studies in answering Singer's queries, he ultimately relies on his deep knowledge of Buddhist teachings to prevail in most of their conversations. The book is a bit unbalanced in this respect, coming off as a pro-meditation (though non-religious) treatise.

The ride through *Beyond the Self* is rapid and enjoyable. Those who practice or are interested in meditation will gain much from reading it, as will those with a passing or professional interest in neuroscience. The dialog style can be discursive, but it also requires that the brain attend to multiple aspects of meditative neuroscience in a way traditionally structured books do not. By example, *Beyond the Self* persuasively suggests that multiple narrative and evidentiary streams are a way to develop contextualized wisdom. That may be a good enough reason to read this fascinating book.

Endnotes

1

Examining the Causes of Autism

1. De Rubeis, S., et al., (2014) Synaptic, transcriptional and chromatin genes disrupted in autism. Nature, 515(7526): p. 209-15.
2. Geschwind, D.H. and M.W. State, (2015) Gene hunting in autism spectrum disorder: on the path to precision medicine. Lancet Neurol, 14(11): p. 1109-20.
3. De Rubeis, S. and J.D. Buxbaum, (2015) Genetics and genomics of autism spectrum disorder: embracing complexity. Hum Mol Genet, 24(R1): p. R24-31.
4. Folstein, S. and M. Rutter, (1977) Infantile autism: a genetic study of 21 twin pairs. J Child Psychol Psychiatry, 18(4): p. 297-321.
5. Hallmayer, J., et al., (2011) Genetic heritability and shared environmental factors among twin pairs with autism. Arch Gen Psychiatry, 68(11): p. 1095-102.
6. Colvert, E., et al., (2015) Heritability of Autism Spectrum Disorder in a UK Population-Based Twin Sample. JAMA Psychiatry, 72(5): p. 415-23.
7. Tick, B., et al., (2016) Autism Spectrum Disorders and Other Mental Health Problems: Exploring Etiological Overlaps and Phenotypic Causal Associations. J Am Acad Child Adolesc Psychiatry, 55(2): p. 106-13 e4.
8. Bourgeron, T., (2016) The genetics and neurobiology of ESSENCE: The third Birgit Olsson lecture. Nord J Psychiatry, 70(1): p. 1-9.
9. Richardson, K. and S. Norgate, (2005) The equal environments assumption of classical twin studies may not hold. Br J Educ Psychol, 75(Pt 3): p. 339-50.
10. Czyz, W., et al., (2012) Genetic, environmental and stochastic factors in monozygotic twin discordance with a focus on epigenetic differences. BMC Med, 10: p. 93.
11. Duszak, R.S., (2009) Congenital rubella syndrome--major review. Optometry, 80(1): p. 36-43.
12. Chess, S., (1971) Autism in children with congenital rubella. J Autism Child Schizophr, 1(1): p. 33-47.
13. Chess, S., (1977) Follow-up report on autism in congenital rubella. J Autism Child Schizophr, 7(1): p. 69-81.
14. Libbey, J.E., et al., (2005) Autistic disorder and viral infections. J Neurovirol, 11(1): p. 1-10.
15. Collier, S.A., et al., (2009) Prevalence of self-reported infection during pregnancy among control mothers in the National Birth Defects Prevention Study. Birth Defects Res A Clin Mol Teratol, 85(3): p. 193-201.
16. Atladottir, H.O., et al., (2010) Maternal infection requiring hospitalization during pregnancy and autism spectrum disorders. J Autism Dev Disord, 40(12): p. 1423-30.

17. Atladottir, H.O., et al., (2012) Autism after infection, febrile episodes, and antibiotic use during pregnancy: an exploratory study. Pediatrics, 130(6): p. e1447-54.
18. Patterson, P.H., (2011) Maternal infection and immune involvement in autism. Trends Mol Med, 17(7): p. 389-94.
19. Shi, L., et al., (2003) Maternal influenza infection causes marked behavioral and pharmacological changes in the offspring. J Neurosci, 23(1): p. 297-302.
20. Zerbo, O., et al., (2013) Is maternal influenza or fever during pregnancy associated with autism or developmental delays? Results from the CHARGE (CHildhood Autism Risks from Genetics and Environment) study. J Autism Dev Disord, 43(1): p. 25-33.
21. Zerbo, O., et al., (2015) Maternal Infection During Pregnancy and Autism Spectrum Disorders. J Autism Dev Disord, 45(12): p. 4015-25.
22. Zerbo, O., et al., (2016) Association Between Influenza Infection and Vaccination During Pregnancy and Risk of Autism Spectrum Disorder. JAMA Pediatr.
23. Schwartzer, J.J., et al., (2016) Behavioral impact of maternal allergic-asthma in two genetically distinct mouse strains. Brain Behav Immun.
24. Schwartzer, J.J., et al., (2013) Maternal immune activation and strain specific interactions in the development of autism-like behaviors in mice. Transl Psychiatry, 3: p. e240.
25. Cooper, G.S., M.L. Bynum, and E.C. Somers, (2009) Recent insights in the epidemiology of autoimmune diseases: improved prevalence estimates and understanding of clustering of diseases. J Autoimmun, 33(3-4): p. 197-207.
26. Goldsmith, C.A. and D.P. Rogers, (2008) The case for autoimmunity in the etiology of schizophrenia. Pharmacotherapy, 28(6): p. 730-41.
27. Dalmau, J., et al., (2008) Anti-NMDA-receptor encephalitis: case series and analysis of the effects of antibodies. Lancet Neurol, 7(12): p. 1091-8.
28. Diamond, B., et al., (2013) Brain-reactive antibodies and disease. Annu Rev Immunol, 31: p. 345-85.
29. Braunschweig, D., et al., (2008) Autism: maternally derived antibodies specific for fetal brain proteins. Neurotoxicology, 29(2): p. 226-31.
30. Fox-Edmiston, E. and J. Van de Water, (2015) Maternal Anti-Fetal Brain IgG Autoantibodies and Autism Spectrum Disorder: Current Knowledge and its Implications for Potential Therapeutics. CNS Drugs, 29(9): p. 715-24.
31. Stromland, K., et al., (1994) Autism in thalidomide embryopathy: a population study. Dev Med Child Neurol, 36(4): p. 351-6.
32. Bromley, R.L., et al., (2013) The prevalence of neurodevelopmental disorders in children prenatally exposed to antiepileptic drugs. J Neurol Neurosurg Psychiatry, 84(6): p. 637-43.
33. Levitt, P., (2011) Serotonin and the autisms: a red flag or a red herring? Arch Gen Psychiatry, 68(11): p. 1093-4.
34. Kaplan, Y.C., et al., (2016) Prenatal selective serotonin reuptake inhibitor use and the risk of autism spectrum disorder in children: A systematic review and meta-analysis. Reprod Toxicol, 66: p. 31-43.
35. Ornoy, A., L. Weinstein-Fudim, and Z. Ergaz, (2015) Prenatal factors associated with autism spectrum disorder (ASD). Reprod Toxicol, 56: p. 155-69.
36. Mandy, W. and M.C. Lai, (2016) Annual Research Review: The role of the environment in the developmental psychopathology of autism spectrum condition. J Child Psychol Psychiatry, 57(3): p. 271-92.

37. Nordahl, C.W., et al., (2011) Brain enlargement is associated with regression in preschool-age boys with autism spectrum disorders. Proc Natl Acad Sci U S A, 108(50): p. 20195-200.

38. Demicheli, V., et al., (2012) Vaccines for measles, mumps and rubella in children. Cochrane Database Syst Rev, (2): p. CD004407.

39. Medicine, I.o., Adverse effects of vaccines: Evidence and Causality, K. Stratton, et al., Editors. 2012.

40. Nelson, C.A., N.A. Fox, and C.H. Zeanah, Romania's abandoned children. 2014, Cambridge MA: Harvard University Press.

41. Rutter, M., et al., (1999) Quasi-autistic patterns following severe early global privation. English and Romanian Adoptees (ERA) Study Team. J Child Psychol Psychiatry, 40(4): p. 537-49.

42. Bernier, R., et al., (2014) Disruptive CHD8 mutations define a subtype of autism early in development. Cell, 158(2): p. 263-76.

2

Next Generation House Call

1. Meyer GS, Gibbons R V. House calls to the elderly--a vanishing practice among physicians. N Engl J Med. 1997;337(25):1815-1820. doi:10.1056/NEJM199712183372507.

2. Topol E. The Patient Will See You Now: The Future of Medicine Is in Your Hands. New York: Basic Books; 2015.

3. Daschle T, Dorsey ER. The return of the house call. Ann Intern Med. 2015;162(8):587-588. doi:10.7326/M14-2769.

4. Howell N. The Doctor's Office of 2024 — 4 Predictions for the Future. The Profitable Practice. http://profitable-practice.softwareadvice.com/doctors-office-of-2024-0514/. Published 2014. Accessed February 6, 2017.

5. Pennic J. Report: Telehealth Video Visits to Reach 158M by 2020. HIT Consultant. http://hitconsultant.net/2015/06/25/report-telehealth-video-visits-to-reach-158m-by-2020/. Published 2015. Accessed February 6, 2017.

6. Mearian L. Almost one in six doctor visits will be virtual this year. Computer World. http://www.computerworld.com/article/2490959/healthcare-it-almost-one-in-six-doctor-visits-will-be-virtual-this-year.html. Published 2014. Accessed February 6, 2017.

7. Dorsey ER, Topol EJ. State of Telehealth. N Engl J Med. 2016;375(2):154-161. doi:10.1056/NEJMra1601705.

8. Preston J, Brown FW, Hartley B. Using telemedicine to improve health care in distant areas. Hosp Community Psychiatry. 1992;43(1):25-32. doi:10.1176/ps.43.1.25.

9. Teich J, Ali MM, Lynch S, Mutter R. Utilization of Mental Health Services by Veterans Living in Rural Areas. J Rural Heal. 2016. doi:10.1111/jrh.12221.

10. Chari KA, Simon AE, DeFrances CJ, Statistics NC for H, Maruschak L, Statistics B of J. National Survey of Prison Health Care: Selected Findings. Natl Health Stat Report. 2016;96:1-23.

11. Brenes GA, Danhauer SC, Lyles MF, Hogan PE, Miller ME. Telephone-Delivered Cognitive Behavioral Therapy and Telephone-Delivered Nondirective Support-

ive Therapy for Rural Older Adults With Generalized Anxiety Disorder: A Randomized Clinical Trial. JAMA Psychiatry. 2015;72(10):1012-1020. doi:10.1001/jamapsychiatry.2015.1154.

12. Huffman JC, Mastromauro CA, Beach SR, et al. Collaborative care for depression and anxiety disorders in patients with recent cardiac events: the Management of Sadness and Anxiety in Cardiology (MOSAIC) randomized clinical trial. JAMA Intern Med. 2014;174(6):927-935. doi:10.1001/jamainternmed.2014.739.

13. Luxton DD, Pruitt LD, Wagner A, Smolenski DJ, Jenkins-Guarnieri MA, Gahm G. Home-based telebehavioral health for U.S. military personnel and veterans with depression: A randomized controlled trial. J Consult Clin Psychol. 2016;84(11):923-934. doi:http://dx.doi.org/10.1037/ccp0000135.

14. Forney JC, Pyne JM, Mouden SB, et al. Practice-based versus telemedicine-based collaborative care for depression in rural federally qualified health centers: a pragmatic randomized comparative effectiveness trial. Am J Psychiatry2. 2013;170(4):414-425. doi:10.1176/appi.ajp.2012.12050696.

15. Fortney JC, Pyne JM, Kimbrell T a., et al. Telemedicine-based collaborative care for posttraumatic stress disorder: a randomized clinical trial. JAMA Psychiatry. 2015;72(1):58-67. doi:http://dx.doi.org/10.1001/jamapsychiatry.2014.1575.

16. Office of Public and Intergovernmental Affairs. VA Telehealth Services Served Over 690,000 Veterans In Fiscal Year 2014. Press Release. https://www.va.gov/opa/pressrel/pressrelease.cfm?id=2646. Published 2014. Accessed February 6, 2017.

17. Hoge CW, Castro CA, Messer SC, McGurk D, Cotting DI, Koffman RL. Combat Duty in Iraq and Afganistan, mental health problems, and barriers to care. N Engl J Med. 2004;351(1):13-22. doi:10.1056/NEJMoa040603.

18. Russo JE, McCool RR, Davies L. VA Telemedicine: An Analysis of Cost and Time Savings. Telemed J E Health. 2016;22(3):209-215. doi:10.1089/tmj.2015.0055.

19. Levine SR, Gorman M. "Telestroke" : the application of telemedicine for stroke. Stroke. 1999;30(2):464-469. doi:10.1161/01.STR.30.2.464.

20. Kepplinger J, Barlinn K, Deckert S, Scheibe M, Bodechtel U, Schmitt J. Safety and efficacy of thrombolysis in telestroke: A systematic review and meta-analysis. Neurology. 2016;87(13):1344-1351. doi:10.1212/WNL.0000000000003148.

21. Nelson RE, Saltzman GM, Skalabrin EJ, Demaerschalk BM, Majersik JJ. The cost-effectiveness of telestroke in the treatment of acute ischemic stroke. Neurology. 2011;77(17):1590-1598. doi:10.1212/WNL.0b013e318234332d.

22. Silva GS, Farrell S, Shandra E, Viswanathan A, Schwamm LH. The status of telestroke in the united states: A survey of currently active stroke telemedicine programs. Stroke. 2012;43(8):2078-2085. doi:10.1161/STROKEAHA.111.645861.

23. Hubert GJ, Muller-Barna P, Audebert HJ. Recent advances in TeleStroke: A systematic review on applications in prehospital management and Stroke Unit treatment or TeleStroke networking in developing countries. Int J Stroke. 2014;9(8):968-973. doi:10.1111/ijs.12394.

24. Muller KI, Alstadhaug KB, Bekkelund SI. Acceptability, feasibility, and cost of telemedicine for nonacute headaches: A randomized study comparing video and traditional consultations. J Med Internet Res. 2016;18(5). doi:10.2196/jmir.5221.

25. Wechsler LR. Advantages and Limitations of Teleneurology. JAMA Neurol. 2015;72(3):349. doi:10.1001/jamaneurol.2014.3844.

26. George BP, Scoglio NJ, Reminick JI, et al. Telemedicine in Leading US Neurology Departments. The Neurohospitalist. 2012;2(4):123-128. doi:10.1177/1941874412450716.

27. Dorsey ER, Venkataraman V, Grana MJ, et al. Randomized controlled clinical trial of "virtual house calls" for Parkinson disease. JAMA Neurol. 2013;70(5):565-570. doi:10.1001/jamaneurol.2013.123.

28. Wilkinson JR, Spindler M, Wood SM, et al. High patient satisfaction with telehealth in Parkinson disease A randomized controlled study. Neurol Clin Pract. 2016;6(3):241-251. doi:10.1212/CPJ.0000000000000252.

29. Dorsey ER, Vlaanderen FP, Engelen LJ, et al. Moving Parkinson care to the home. Mov Disord. 2016;31(9):1258-1262. doi:10.1002/mds.26744.

30. NIMH. U.S. Leading Categories of Diseases/Disorders. https://www.nimh.nih.gov/health/statistics/disability/us-leading-categories-of-diseases-disorders.shtml. Accessed February 7, 2017.

31. NIMH. Global Leading Categories of Diseases/Disorders. https://www.nimh.nih.gov/health/statistics/global/global-leading-categories-of-diseases-disorders.shtml. Accessed February 7, 2017.

32. Blumberg SJ, Bramlett MD, Kogan MD, Schieve LA, Jones JR, Lu MC. National Health Statistics Reports Number 65 March 20. 2013. 2007.

33. Hebert LE, Scherr PA, Bienias JL, Bennett DA, Evans DA. Alzheimer Disease in the US Population. Arch Neurol. 2003;60(8):1119. doi:10.1001/archneur.60.8.1119.

34. Dorsey ER, Constantinescu R, Thompson JP, et al. Projected number of people with Parkinson disease in the most populous nations, 2005 through 2030. Neurology. 2007;68(5):384-386. doi:10.1212/01.wnl.0000247740.47667.03.

35. Hunt G, Greene R, Whiting C, et al. Caregiving in the U.S. – 2015 Report.; 2015.

36. Gros DF, Lancaster CL, López CM, Acierno R. Treatment satisfaction of home-based telehealth versus in-person delivery of prolonged exposure for combat-related PTSD in veterans. J Telemed Telecare. September 2016:1357633X1667109. doi:10.1177/1357633X16671096.

37. Dorsey ER, Achey MA, Beck CA, et al. National Randomized Controlled Trial of Virtual House Calls for People with Parkinson's Disease: Interest and Barriers. Telemed e-Health. 2016;22(7):590-598. doi:10.1089/tmj.2015.0191.

38. Mair F, Whitten P. Systematic review of studies of patient satisfaction with telemedicine. BMJ. 2000;320(7248).

39. Hyler SE, Gangure DP, Batchelder ST. Can telepsychiatry replace in-person psychiatric assessments? A review and meta-analysis of comparison studies. CNS Spectr. 2005;10(5):403-413. http://www.ncbi.nlm.nih.gov/pubmed/15858458. Accessed March 8, 2017.

40. Anand BN. The Content Trap : A Strategist's Guide to Digital Change. 1st ed. New York: Random House; 2016.

41. Irfan A. The AMA, Telemedicine's Reluctant Advocate. Telemedicine Magazine. http://www.telemedmag.com/features/2016/9/27/the-ama-telemedicines-reluctant-advocate. Accessed February 7, 2017.

42. Oliver DP, Demiris G, Wittenberg-Lyles E, Washington K, Day T, Novak H. A Systematic Review of the Evidence Base for Telehospice. Telemed e-Health. 2012;18(1):38-47. doi:10.1089/tmj.2011.0061.

43. ATA. State Policy Resource Center. http://www.americantelemed.org/main/policy-page/state-policy-resource-center. Published 2017. Accessed February 7, 2017.

44. Neufeld JD, Doarn CR. Telemedicine Spending by Medicare: A Snapshot from 2012. Telemed e-Health. 2015;21(8):686-693. doi:10.1089/tmj.2014.0185.

45. Telehealth Services.; 2016.

46. Chaudhry HJ, Robin LA, Fish EM, Polk DH, Gifford JD. Improving Access and Mobility — The Interstate Medical Licensure Compact. N Engl J Med. 2015;372(17):1581-1583. doi:10.1056/NEJMp1502639.

47. State Laws and Reimbursement Policies. http://www.cchpca.org/state-laws-and-reimbursement-policies. Accessed February 9, 2017.

48. Norris P. Digital Divide : Civic Engagement, Information Poverty, and the Internet Worldwide. New York: Cambridge University Press; 2001.

49. Rainie L. Digital Divides 2015. Pew Research Center: Internet, Science & Tech. http://www.pewinternet.org/2015/09/22/digital-divides-2015/. Published 2015. Accessed February 7, 2017.

50. Fox S, Purcell K. Chronic Disease and the Internet. Pew Research Center: Internet, Science & Tech. http://www.pewinternet.org/2010/03/24/chronic-disease-and-the-internet/. Published 2010. Accessed February 7, 2017.

51. Kurzweil R. The Singularity Is near : When Humans Transcend Biology. London: Penguin Group; 2005.

52. O'Gorman LD, Hogenbirk JC, Warry W. Clinical Telemedicine Utilization in Ontario over the Ontario Telemedicine Network. Telemed J E Health. 2016;22(6):473-479. doi:10.1089/tmj.2015.0166.

53. Toffler A. Future Shock. Bantam Books; 1990.

54. Pearl R. Kaiser Permanente Northern California: current experiences with internet, mobile, and video technologies. Health Aff (Millwood). 2014;33(2):251-257. doi:10.1377/hlthaff.2013.1005.

55. Northern California: Current Experiences with Internet, Mobile, and Video Technologies.; 2016.

56. Willis AW, Schootman M, Tran R, et al. Neurologist-associated reduction in PD-related hospitalizations and health care expenditures. Neurology. 2012;79(17):1774-1780. doi:10.1212/WNL.0b013e3182703f92.

57. National Ambulatory Medical Care Survey: 2013 State and National Summary Tables.; 2013.

58. Ray KN, Chari A V., Engberg J, et al. Disparities in Time Spent Seeking Medical Care in the United States. JAMA Intern Med. 2015;175(12):1983. doi:10.1001/jamainternmed.2015.4468.

59. Nassauer S, Davidson K. Wal-Mart Makes Rare Retreat on Home Turf. The Wall Street Journal. https://www.wsj.com/articles/wal-mart-to-close-269-stores-globally-1452868122. Published 2016. Accessed February 7, 2017.

60. Beilfuss L. Macy's to Close About 15% of Stores, Cut Jobs. The Wall Street Journal. https://www.wsj.com/articles/macys-to-close-about-15-of-stores-cut-jobs-1470918685. Published 2016. Accessed February 7, 2017.

61. Fung E. Vacancy Rates Rise at Shopping Centers. The Wall Street Journal. https://www.wsj.com/articles/vacancies-hurt-retail-property-market-1483677002. Published 2017. Accessed February 7, 2017.

62. Phillips E, Smith J. UPS, FedEx Struggle to Keep Up With Surge in Holiday Orders. The Wall Street Journal. https://www.wsj.com/articles/ups-fedex-struggle-to-keep-up-with-surge-in-holiday-orders-1481630401. Published 2016. Accessed February 7, 2017.

3
The Four Pillars of Alzheimer's Prevention

1. Alzheimer's Association, "2016 Alzheimer's Disease Facts and Figures," Alzheimer's & Dementia (2016);12(4). https://www.alz.org/documents_custom/2016-facts-and-figures.pdf.

2. National Institute on Aging, "Number of Alzheimer's Deaths Found to be Underreported", (May 22, 2014). https://www.nia.nih.gov/research/announcements/2014/05/number-alzheimers-deaths-found-be-underreported.

3. National Institutes on Health, "Estimates of Funding for Various Research, Condition, and Disease Categories (RCDC)," (February 10, 2016). https://health.gov/dietaryguidelines/2015/guidelines/chapter-2/current-eating-patterns-in-the-united-states/.

4. T. Ngandu, J. Lehtisalo, A. Solomon, M. Kivipelto, et al, "The Finnish Geriatric Intervention Study to Prevent Cognitive Impairment and Disability (FINGER): Study Design and Progress," The Lancet (2015);385(9984): 2255–2263.
doi: http://dx.doi.org/10.1016/S0140-6736(15)60461-5.
http://www.thelancet.com/journals/lancet/article/PIIS0140-6736(15)60461-5/abstract.

5. S. Andrieu, B. Vellas, "MAPT Study: A Multidomain Approach for Preventing Alzheimer's Disease," (2016). Presentation of soon to be published data at ISTA-ART, Alzheimer's Association International Conference, Toronto, Canada.

6. US Office of Disease Prevention and Health Promotion, "Current Eating Patterns in the United States," (2015). https://health.gov/dietaryguidelines/2015/guidelines/chapter-2/current-eating-patterns-in-the-united-states/.

7. B. Shakersain, G. Santoni, S.C. Larsson, W. Xu, "Prudent Diet May Attenuate the Adverse Effects of Western Diet on Cognitive Decline," Alzheimer's & Dementia (2016): 100-09.
doi: 10.1016/j.jalz.2015.08.002.
https://www.researchgate.net/publication/281588866_Prudent_diet_may_attenuate_the_adverse_effects_of_Western_diet_on_cognitive_decline.

8. D.A. Merrill, P. Siddarth, C.A. Raji, N.D. Emerson, F. Rueda, L.M. Ercoli, K.J. Miller, H. Lavretsky, L.M. Harris, A.C. Burggren, S.Y. Bookheimer, J.R. Barrio, G.W. Small, "Modifiable Risk Factors and Brain Positron Emission Tomography Measures of Amyloid and Tau in Nondemented Adults with Memory Complaints," The American Journal of Geriatric Psychiatry (2016): 729-737.
doi: 10.1016/j.jagp.2016.05.007.
https://www.ncbi.nlm.nih.gov/pubmed/27421618/.

9. S.C. Staubo, J.A. Aakre, P. Vemuri, J.A. Syrjanen, M.M. Mielke, Y.E. Geda, W.K. Kremers, M.M. Machulda, D.S. Knopman, R.C. Petersen, C.R. Jack Jr., R.O. Roberts, "Mediterranean Diet, Micronutrients and Macronutrients, and MRI Measures of Cortical Thickness," Alzheimer's & Dementia (2016), published online.

doi: http://dx.doi.org/10.1016/j.jalz.2016.06.2359.
http://www.alzheimersanddementia.com/article/S1552-5260(16)32661-9/pdf.

10. M.C. Morris, C. C. Tangney, Y. Wang, L. L. Barnes, D. Bennett, N. Aggarwal, "MIND Diet Score More Predictive than DASH or Mediterranean Diet Scores," Alzheimer's & Dementia (2014):166.
doi: http://dx.doi.org/10.1016/j.jalz.2014.04.164.
http://www.alzheimersanddementia.com/article/S1552-5260 (14)00292-1/fulltext.

11. A.C. Pereira, D.E. Huddleston, A.M. Brickman, A.A. Sosunov, R. Hen, G.M. McKhann, R. Sloan, F.H. Gage, T.R. Brown, S.A. Small, "An In Vivo Correlate of Exercise-induced Neurogenesis in the Adult Dentate Gyrus," Proceedings of the National Academy of Sciences (2007); 104(13):5638–5643.
doi: 10.1073/pnas.0611721104.
http://www.pnas.org/content/104/13/5638.short.

12. S.J. Lupien, B. McEwen, M. Gunner, et al, "Effects of Stress Throughout the Lifespan on the Brain, Behavior and Cognition," Nature (2009); 10:434-445.
doi:10.1038/nrn2639.
http://www.nature.com/nrn/journal/v10/n6/abs/nrn2639.html.

13. D.S. Khalsa, "Stress, Meditation and Alzheimer's Disease Prevention: Where The Evidence Stands," Journal of Alzheimer's Disease (2015); 48:1-12.
doi 10.3233/JAD-142766.
https://www.ncbi.nlm.nih.gov/pmc/articles/PMC4923750/.

14. A.B. Newberg, N. Wintering, D.S. Khalsa, H. Roggenkamp, M.R. Waldman, "Meditation Effects on Cognitive Function and Cerebral Blood Flow in Subjects with Memory Loss: A Preliminary Study," Journal of Alzheimer's Disease (2010); 20(2):517-26.
doi: 10.3233/JAD-2010-1391.
http://www.ncbi.nlm.nih.gov/pubmed/20164557.

15. T.M. Harrison, S. Weintraub, M.M. Mesulam, and E. Rogalski, "Superior Memory and Higher Cortical Volumes in Unusually Successful Cognitive Aging', Journal of the International Neuropsychological Society (2012);18(6):1081–1085.
doi: 10.1017/S1355617712000847.
https://www.ncbi.nlm.nih.gov/pmc/articles/PMC3547607/.

16. A.S. Moss, N. Wintering, H. Roggenkamp, D.S. Khalsa, M.R. Waldman, D. Monti, A.B. Newberg, "Effects of An 8-Week Meditation Program on Mood and Anxiety in Patients With Memory Loss," Journal of Alternative and Complementary Medicine (2012);18(1):48-53.
doi: 10.1089/acm.2011.0051.
http://www.ncbi.nlm.nih.gov/pubmed/22268968.

17. H. Lavretsky, E. Epel, P. Siddarth, N. Nazarian, N. St. Cyr, D.S. Khalsa, J. Lin, E. Blackburn, M.R. Irwin, "A Pilot Study of Yogic Meditation for Family Dementia Caregivers with Depressive Symptoms: Effects on Mental Health, Cognition, and Telomerase Activity," International Journal of Geriatric Psychiatry (2013);28(1):57-65.
doi: 10.1002/gps.3790.
http://www.ncbi.nlm.nih.gov/pubmed/22407663.

18. K. Innes, T. K. Selfe, D. S. Khalsa, S. Kandati, "Effects of Meditation versus Music Listening on Perceived Stress, Mood, Sleep, and Quality of Life in Adults with

Early Memory Loss: A Pilot Randomized Controlled Trial," Journal of Alzheimer's Disease (2016) 8;52(4):1277-98.

doi: 10.3233/JAD-151106.

http://www.ncbi.nlm.nih.gov/pubmed/27079708.

19. H.A. Eyre, B. Acevedo, H. Yang, P. Siddarth, K. Van Dyk, L. Ercoli, A. M. Leaver, N. St.Cyr, K. Narr, B.T. Baune, D.S. Khalsa, H. Lavretsky, "Changes in Neural Connectivity and Memory Following a Yoga Intervention for Older Adults: A Pilot Study," Journal of Alzheimer's Disease (2016); 52(2): 673–684.

doi: 10.3233/JAD-150653.

http://www.ncbi.nlm.nih.gov/pubmed/27060939.

20. Y. Yang, A. M. Leaver, P. S. Pattharee Paholpak, L. Ercoli, N. M. St. Cyr, H. A. Eyre, K. L. Narr, D. S. Khalsa, H. Lavretsky, "Neurochemical and Neuroanatomical Plasticity Following Memory Training and Yoga Interventions in Older Adults with Mild Cognitive Impairment," Frontiers in Aging Neuroscience (2016); 8: 277.

http://dx.doi.org/10.3389/fnagi.2016.00277.

21. C.D. Ryff, B.H. Singer, L. G. Dienberg, "Positive Health: Connecting Well-Being with Biology," Philosophical Transactions of the Royal Society of London, Series B, Biological Sciences (2004);359: 1383-1394.

doi:10.1098/rstb.2004.1521.

http://scholar.google.com/citations?view_op=view_citation&hl=en&user=Y-Ilu3jYAAAAJ&citation_for_view=YIlu3jYAAAAJ:WF5omc3nYNoC.

22. P.A. Boyle, A.S. Buchman, L.L. Barnes, D.A. Bennett, "Effect of Purpose in Life on Incident Alzheimer's Disease and Mild Cognitive Impairment in Community-Dwelling Older Persons," Archives of General Psychiatry (2010);67(3): 304-310.

doi:10.1001/archgenpsychiatry.2009.208.

http://jamanetwork.com/journals/jamapsychiatry/fullarticle/210648.

23. Y. Kaufman, D. Anaki, M. Binns, M. Freedman, "Cognitive decline in Alzheimer Disease: Impact of Spirituality, Religiosity, and QOL," Neurology (2007)1;68(18):1509-14.

doi: 10.1212/01.wnl.0000260697.66617.59.

https://www.ncbi.nlm.nih.gov/pubmed/17470754.

24. H. Lavretsky, "Resilience and Aging Research and Practice," (2014):94-129. Johns Hopkins University Press, Baltimore, Maryland.

25. K. Sheardova, Z. Nedleska, R. Sumec, R. Marciniak, S. Belaskova, M. Uller, J. Hort, "The Effect of Spirituality/Religiosity on Regional Brain Atrophy in Subjects at Risk of Alzheimer's Disease, 3-year Follow-up Data from Czech Brain Aging Study," Personal Communication. To be presented at AAIC July 2017.

4

Gut Feelings on Parkinson's and Depression

1. Stilling RM, Dinan TG, Cryan JF. Microbial genes, brain & behaviour - epigenetic regulation of the gut-brain axis. Genes Brain Behav 2014, 13(1): 69-86.

2. Donaldson IM. James Parkinson's essay on the shaking palsy. J R Coll Physicians Edinb 2015, 45(1): 84-86.

3. Jagmag SA, Tripathi N, Shukla SD, Maiti S, Khurana S. Evaluation of Models of

Parkinson's Disease. Front Neurosci 2015, 9: 503.

4. Moloney RD, Desbonnet L, Clarke G, Dinan TG, Cryan JF. The microbiome: stress, health and disease. Mamm Genome 2014, 25(1-2): 49-74.

5. Grenham S, Clarke G, Cryan JF, Dinan TG. Brain-gut-microbe communication in health and disease. Front Physiol 2011, 2: 94.

6. Phillips RJ, Walter GC, Wilder SL, Baronowsky EA, Powley TL. Alpha-synuclein-immunopositive myenteric neurons and vagal preganglionic terminals: autonomic pathway implicated in Parkinson's disease? Neuroscience 2008, 153(3): 733-750.

7. Svensson E, Horvath-Puho E, Thomsen RW, Djurhuus JC, Pedersen L, Borghammer P, et al. Does vagotomy reduce the risk of Parkinson's disease: The authors reply. Ann Neurol 2015, 78(6): 1012-1013.

8. Scheperjans F, Aho V, Pereira PA, Koskinen K, Paulin L, Pekkonen E, et al. Gut microbiota are related to Parkinson's disease and clinical phenotype. Mov Disord 2015, 30(3): 350-358.

9. Sampson TR, Debelius JW, Thron T, Janssen S, Shastri GG, Ilhan ZE, et al. Gut Microbiota Regulate Motor Deficits and Neuroinflammation in a Model of Parkinson's Disease. Cell 2016, 167(6): 1469-1480 e1412.

10. Buddhala C, Loftin SK, Kuley BM, Cairns NJ, Campbell MC, Perlmutter JS, et al. Dopaminergic, serotonergic, and noradrenergic deficits in Parkinson disease. Ann Clin Transl Neurol 2015, 2(10): 949-959.

11. Jiang H, Ling Z, Zhang Y, Mao H, Ma Z, Yin Y, et al. Altered fecal microbiota composition in patients with major depressive disorder. Brain Behav Immun 2015, 48: 186-194.

12. Kelly JR, Borre Y, C OB, Patterson E, El Aidy S, Deane J, et al. Transferring the blues: Depression-associated gut microbiota induces neurobehavioural changes in the rat. J Psychiatr Res 2016, 82: 109-118.

13. Borody TJ, Brandt LJ, Paramsothy S, Agrawal G. Fecal microbiota transplantation: a new standard treatment option for Clostridium difficile infection. Expert Rev Anti Infect Ther 2013, 11(5): 447-449.

14. Allen AP, Hutch W, Borre YE, Kennedy PJ, Temko A, Boylan G, et al. Bifidobacterium longum 1714 as a translational psychobiotic: modulation of stress, electrophysiology and neurocognition in healthy volunteers. Transl Psychiatry 2016, 6(11): e939.

5

Genetics and ALS: Cause for Optimism

1. Goetz CG. Amyotrophic Lateral Sclerosis: Early Contributions of Jean-Martin Charcot. Muscle Nerve 2000;23:336–343.

2. Peters OM, Ghasemi M, Brown RH Jr. Emerging mechanisms of molecular pathology in ALS. J Clin Invest. 2015 May;125(5):1767-79.

3. Brain, WR. Diseases of the nervous system. Oxford University Press; Oxford: 1962. Motor Neuron Disease; 531-43.

4. Miller RG1, Mitchell JD, Moore DH. Riluzole for amyotrophic lateral sclerosis (ALS)/motor neuron disease (MND). Cochrane Database Syst Rev. 2012 Mar 14;(3):CD001447.

5. Chen L, Liu X, Tang L, Zhang N, Fan D. Long-Term Use of Riluzole Could

Improve the Prognosis of Sporadic Amyotrophic Lateral Sclerosis Patients: A Real-World Cohort Study in China. Front Aging Neurosci. 2016 Oct 24;8:246.

6. Abe K, Itoyama Y, Sobue G, Tsuji S, Aoki M, Doyu M, Hamada C, Kondo K, Yoneoka T, Akimoto M, Yoshino H; Edaravone ALS Study Group. Confirmatory double-blind, parallel-group, placebo-controlled study of efficacy and safety of edaravone (MCI-186) in amyotrophic lateral sclerosis patients. Amyotroph Lateral Scler Frontotemporal Degener. 2014 Dec;15(7-8):610-7.

7. Watanabe T, Yuki S, Egawa M, Nishi H. Protective effects of MCI-186 on cerebral ischemia: possible involvement of free radical scavenging and antioxidant actions. J Pharmacol Exp Ther. 1994 Mar;268(3):1597-604.

8. Rosen, D. R. et al. Mutations in Cu/Zn superoxide dismutase gene are associated with familial amyotrophic lateral sclerosis. Nature 362, 59–62 (1993).

9. Andersen PM, Sims KB, Xin WW, Kiely R, O'Neill G, Ravits J, Pioro E, Harati Y, Brower RD, Levine JS, Heinicke HU, Seltzer W, Boss M, Brown RH Jr. Sixteen novel mutations in the Cu/Zn superoxide dismutase gene in amyotrophic lateral sclerosis: a decade of discoveries, defects and disputes. Amyotroph Lateral Scler Other Motor Neuron Disord. 2003 Jun;4(2):62-73.

10. Bruijn, L. I. et al. ALS-linked SOD1 mutant G85R mediates damage to astrocytes and promotes rapidly progressive disease with SOD1-containing inclusions. Neuron 18, 327–338 (1997)

11. International Human Genome Sequencing Consortium. Initial sequencing and analysis of the human genome. Nature. 2001;409:860–921.

12. International Human Genome Sequencing Consortium. Finishing the euchromatic sequence of the human genome. Nature. 2004;431:931–45

13. https://www.theguardian.com/society/2015/may/30/als-after-the-ice-bucket-challenge

14. J. Paul Taylor, Robert H. Brown Jr & Don W. Cleveland. Decoding ALS: from genes to mechanism, Nature. 2016 Nov 10;539(7628):197-206

15. Renton AE, Chiò A, Traynor BJ. State of play in amyotrophic lateral sclerosis genetics. Nat Neurosci. 2014 Jan;17(1):17-23

16. Al-Chalabi A, van den Berg LH, Veldink J. Gene discovery in amyotrophic lateral sclerosis: implications for clinical management. Nat Rev Neurol. 2016 Dec 16.

17. D. G. MacArthur, T. A. Manolio, D. P. Dimmock, H. L. Rehm, J. Shendure, G. R. Abecasis, D. R. Adams, R. B. Altman, S. E. Antonarakis, E. A. Ashley, J. C. Barrett, L. G. Biesecker, D. F. Conrad, G. M. Cooper, N. J. Cox, M. J. Daly, M. B. Gerstein, D. B. Goldstein, J. N. Hirschhorn, S. M. Leal, L. A. Pennacchio, J. A. Stamatoyannopoulos, S. R. Sunyaev, D. Valle, B. F. Voight, W. Winckler and C. Gunter. Guidelines for investigating causality of sequence variants in human disease. Nature. 2014 Apr 24; 508(7497): 469–476.

18. Cady J, Allred P, Bali T, Pestronk A, Goate A, Miller TM, Mitra RD, Ravits J, Harms MB, Baloh RH. Amyotrophic lateral sclerosis onset is influenced by the burden of rare variants in known amyotrophic lateral sclerosis genes. Ann Neurol. 2015 Jan;77(1):100-13.

19. Harari O, Cruchaga C. Paving the road for the study of epigenetics in neurodegenerative diseases. Acta Neuropathol. 2016 Oct;132(4):483-5.

20. Shatunov A, Mok K, Newhouse S, Weale ME, Smith B, Vance C, Johnson L, Veldink JH, van Es MA, van den Berg LH, Robberecht W, Van Damme P, Hardiman O, Farmer AE, Lewis CM, Butler AW, Abel O, Andersen PM, Fogh I, Silani V,

Chiò A, Traynor BJ, Melki J, Meininger V, Landers JE, McGuffin P, Glass JD, Pall H, Leigh PN, Hardy J, Brown RH Jr, Powell JF, Orrell RW, Morrison KE, Shaw PJ, Shaw CE, Al-Chalabi A. Chromosome 9p21 in sporadic amyotrophic lateral sclerosis in the UK and seven other countries: a genome-wide association study. Lancet Neurol. 2010 Oct;9(10):986-94.

21. Benatar M, Stanislaw C, Reyes E, Hussain S, Cooley A, Fernandez MC, Dauphin DD, Michon SC, Andersen PM, Wuu J. Presymptomatic ALS genetic counseling and testing: Experience and recommendations. Neurology. 2016 Jun 14;86(24):2295-302

22. Warr A, Robert C, Hume D, Archibald A, Deeb N, Watson M. Exome Sequencing: Current and Future Perspectives. G3 (Bethesda). 2015 Jul 2;5(8):1543-50.

23. Elizabeth T. Cirulli, Brittany N. Lasseigne. Exome sequencing in amyotrophic lateral sclerosis identifies risk genes and pathways. Science. 2015 Mar 27; 347(6229): 1436–1441.

24. Pottier C, Bieniek KF, Finch N, van de Vorst M, Baker M, Perkersen R, Brown P, Ravenscroft T, van Blitterswijk M, Nicholson AM, DeTure M, Knopman DS, Josephs KA, Parisi JE, Petersen RC, Boylan KB, Boeve BF, Graff-Radford NR, Veltman JA, Gilissen C, Murray ME, Dickson DW, Rademakers R. Whole-genome sequencing reveals important role for TBK1 and OPTN mutations in frontotemporal lobar degeneration without motor neuron disease. Acta Neuropathol. 2015 Jul;130(1):77-92

25. Therrien M, Dion PA, Rouleau GA. ALS: Recent Developments from Genetics Studies. Curr Neurol Neurosci Rep. 2016 Jun;16(6):59.

26. Evers MM, Toonen LJ, van Roon-Mom WM. Antisense oligonucleotides in therapy for neurodegenerative disorders. Adv Drug Deliv Rev. 2015 Jun 29;87:90-103. Review.

27. Stephenson ML, Zamecnik PC. Inhibition of Rous sarcoma viral RNA translation by a specific oligodeoxyribonucleotide. Proc Natl Acad Sci U S A. 1978 Jan;75(1):285-8.

28. Smith, R.A., Miller, T.M., Yamanaka, K., Monia, B.P., Condon, T.P., Hung, G., Lobsiger, C.S., Ward, C.M., McAlonis-Downes, M., Wei, H., Wancewicz EV, Bennett CF, Cleveland DW. (2006). Antisense oligonucleotide therapy for neurodegenerative disease. J. Clin. Invest. 116, 2290–2296.

29. Miller TM, Pestronk A, David W, Rothstein J, Simpson E, Appel SH, Andres PL, Mahoney K, Allred P, Alexander K, Ostrow LW, Schoenfeld D, Macklin EA, Norris DA, Manousakis G, Crisp M, Smith R, Bennett CF, Bishop KM, Cudkowicz ME. An antisense oligonucleotide against SOD1 delivered intrathecally for patients with SOD1 familial amyotrophic lateral sclerosis: a phase 1, randomised, first-in-man study. Lancet Neurol. 2013 May;12(5):435-42.

30. Jiang J, Zhu Q, Gendron TF, Saberi S, McAlonis-Downes M, Seelman A, Stauffer JE, Jafar-Nejad P, Drenner K, Schulte D, Chun S, Sun S, Ling SC, Myers B, Engelhardt J, Katz M, Baughn M, Platoshyn O, Marsala M, Watt A, Heyser CJ, Ard MC, De Muynck L, Daughrity LM, Swing DA, Tessarollo L, Jung CJ, Delpoux A, Utzschneider DT, Hedrick SM, de Jong PJ, Edbauer D, Van Damme P, Petrucelli L, Shaw CE, Bennett CF, Da Cruz S, Ravits J, Rigo F, Cleveland DW, Lagier-Tourenne C. Gain of Toxicity from ALS/FTD-Linked Repeat Expansions in C9ORF72 Is Alleviated by Antisense Oligonucleotides Targeting GGG-GCC-Containing RNAs. Neuron. 2016 May 4;90(3):535-50.

31. Terry SF. The study is open: Participants are now recruiting investigators. Sci Transl Med. 2017 Jan 4;9(371).
32. http://www.fda.gov/downloads/MedicalDevices/DeviceRegulationandGuidance/GuidanceDocuments/UCM509837.pdf

6

The Sleeping Brain

1. Cirelli, C. & Tononi, G. Is sleep essential? PLoS Biol 6, e216 (2008).
2. Borbely, A.A., Daan, S., Wirz-Justice, A. & Deboer, T. The two-process model of sleep regulation: a reappraisal. J Sleep Res 25, 131-143 (2016).
3. Mullington, J.M., Simpson, N.S., Meier-Ewert, H.K. & Haack, M. Sleep loss and inflammation. Best Pract Res Clin Endocrinol Metab 24, 775-784 (2010).
4. Arble, D.M., et al. Impact of Sleep and Circadian Disruption on Energy Balance and Diabetes: A Summary of Workshop Discussions. Sleep 38, 1849-1860 (2015).
5. Lo, J.C., et al. Effects of Partial and Acute Total Sleep Deprivation on Performance across Cognitive Domains, Individuals and Circadian Phase. PloS one 7, e45987 (2012).
6. Rupp, T.L., Wesensten, N.J. & Balkin, T.J. Trait-like vulnerability to total and partial sleep loss. Sleep 35, 1163-1172 (2012).
7. Goel, N., Basner, M., Rao, H. & Dinges, D.F. Circadian rhythms, sleep deprivation, and human performance. Prog Mol Biol Transl Sci 119, 155-190 (2013).
8. Killgore, W.D. Effects of sleep deprivation on cognition. Progress in brain research 185, 105-129 (2010).
9. Banks, S. & Dinges, D.F. Behavioral and physiological consequences of sleep restriction. J Clin Sleep Med 3, 519-528 (2007).
10. Mednick, S.C., et al. The restorative effect of naps on perceptual deterioration. Nature neuroscience 5, 677-681 (2002).
11. Ficca, G., Axelsson, J., Mollicone, D.J., Muto, V. & Vitiello, M.V. Naps, cognition and performance. Sleep Med Rev 14, 249-258 (2010).
12. Tononi, G. & Cirelli, C. Sleep and synaptic homeostasis: a hypothesis. Brain research bulletin 62, 143-150 (2003).
13. Tononi, G. & Cirelli, C. Sleep function and synaptic homeostasis. Sleep Med Rev 10, 49-62 (2006).
14. Tononi, G. & Cirelli, C. Sleep and the price of plasticity: from synaptic and cellular homeostasis to memory consolidation and integration. Neuron 81, 12-34 (2014).
15. Attwell, D. & Gibb, A. Neuroenergetics and the kinetic design of excitatory synapses. Nat Rev Neurosci 6, 841-849 (2005).
16. Hallermann, S., de Kock, C.P., Stuart, G.J. & Kole, M.H. State and location dependence of action potential metabolic cost in cortical pyramidal neurons. Nature neuroscience 15, 1007-1014 (2012).
17. Buzsaki, G. & Mizuseki, K. The log-dynamic brain: how skewed distributions affect network operations. Nat Rev Neurosci 15, 264-278 (2014).
18. Suthana, N. & Fried, I. Percepts to recollections: insights from single neuron recordings in the human brain. Trends in cognitive sciences 16, 427-436 (2012).
19. Vyazovskiy, V.V., Cirelli, C., Pfister-Genskow, M., Faraguna, U. & Tononi, G. Molecular and electrophysiological evidence for net synaptic potentiation in wake

and depression in sleep. Nature neuroscience 11, 200-208 (2008).

20. Huber, R., et al. Human cortical excitability increases with time awake. Cerebral cortex 23, 332-338 (2013).

21. Cirelli, C. Sleep, synaptic homeostasis and neuronal firing rates. Curr Opin Neurobiol 44, 72-79 (2017).

22. Liu, Z.W., Faraguna, U., Cirelli, C., Tononi, G. & Gao, X.B. Direct evidence for wake-related increases and sleep-related decreases in synaptic strength in rodent cortex. The Journal of neuroscience : the official journal of the Society for Neuroscience 30, 8671-8675 (2010).

23. Gilestro, G.F., Tononi, G. & Cirelli, C. Widespread changes in synaptic markers as a function of sleep and wakefulness in Drosophila. Science 324, 109-112 (2009).

24. Bushey, D., Tononi, G. & Cirelli, C. Sleep and synaptic homeostasis: structural evidence in Drosophila. Science 332, 1576-1581 (2011).

25. Maret, S., Faraguna, U., Nelson, A.B., Cirelli, C. & Tononi, G. Sleep and waking modulate spine turnover in the adolescent mouse cortex. Nature neuroscience 14, 1418-1420 (2011).

26. de Vivo, L., et al. Ultrastructural evidence for synaptic scaling across the wake/sleep cycle. Science 355, 507-510 (2017).

27. Diering, G.H., et al. Homer1a drives homeostatic scaling-down of excitatory synapses during sleep. Science 355, 511-515 (2017).

28. Olcese, U., Esser, S.K. & Tononi, G. Sleep and synaptic renormalization: a computational study. Journal of neurophysiology 104, 3476-3493 (2010).

29. Nere, A., Hashmi, A., Cirelli, C. & Tononi, G. Sleep Dependent Synaptic Down-Selection (I): Modeling the Benefits of Sleep on Memory Consolidation and Integration. Frontiers in Sleep and Chronobiology 4, 143 (2013).

30. Hashmi, A., Nere, A. & Tononi, G. Sleep Dependent Synaptic Down-Selection (II): Single Neuron Level Benefits for Matching, Selectivity, and Specificity. Frontiers in Sleep and Chronobiology 4, 144 (2013).

31. Billeh, Y.N., et al. Effects of chronic sleep restriction during early adolescence on the adult pattern of connectivity of mouse secondary motor cortex. eneuro, ENEURO. 0053-0016.2016 (2016).

32. Cirelli, C. & Tononi, G. Cortical development, electroencephalogram rhythms, and the sleep/wake cycle. Biol Psychiatry 77, 1071-1078 (2015).

7

The Brain's Emotional Development

1. Ochsner KN, Silvers JA, Buhle JT. Functional imaging studies of emotion regulation: a synthetic review and evolving model of the cognitive control of emotion. Annals of the New York Academy of Sciences. 2012;1251:E1-24.

2. McRae K, Hughes B, Chopra S, Gabrieli JD, Gross JJ, Ochsner KN. The neural bases of distraction and reappraisal. J Cogn Neurosci. Feb;22(2):248-262.

3. Gross JJ. Antecedent- and response-focused emotion regulation: divergent consequences for experience, expression, and physiology. J Pers Soc Psychol. Jan 1998;74(1):224-237.

4. Cisler JM, Olatunji BO. Emotion Regulation and Anxiety Disorders. Current psychiatry reports. 2012;14:182-187.

5. Campos JJ, Frankel CB, Camras L. On the nature of emotion regulation. Child Dev. Mar-Apr 2004;75(2):377-394.

6. Thompson JL, Nelson AJ. Middle childhood and modern human origins. Human Nature. Sep 2011;22(3):249-280.

7. Werker JF, Hensch TK. Critical Periods in Speech Perception: New Directions. http://dx.doi.org/10.1146/annurev-psych-010814-015104. 2015.

8. Quirk GJ, Beer JS. Prefrontal involvement in the regulation of emotion: convergence of rat and human studies. Current Opinion in Neurobiology. Dec 2006;16(6):723-727.

9. Hare TA, Tottenham N, Davidson MC, Glover GH, Casey BJ. Contributions of amygdala and striatal activity in emotion regulation. Biological Psychiatry. Mar 15 2005;57(6):624-632.

10. Sokol-Hessner P, Camerer CF, Phelps EA. Emotion regulation reduces loss aversion and decreases amygdala responses to losses. Soc Cogn Affect Neurosci. Mar 2013;8(3):341-350.

11. Barbas H, Ghashghaei H, Dombrowski SM, Rempel-Clower NL. Medial prefrontal cortices are unified by common connections with superior temporal cortices and distinguished by input from memory-related areas in the rhesus monkey. J Comp Neurol. Aug 2 1999;410(3):343-367.

12. Ghashghaei HT, Barbas H. Pathways for emotion: interactions of prefrontal and anterior temporal pathways in the amygdala of the rhesus monkey. Neuroscience. 2002;115(4):1261-1279.

13. Ghashghaei HT, Hilgetag CC, Barbas H. Sequence of information processing for emotions based on the anatomic dialogue between prefrontal cortex and amygdala. Neuroimage. Feb 1 2007;34(3):905-923.

14. Cho YT, Ernst M, Fudge JL. Cortico-amygdala-striatal circuits are organized as hierarchical subsystems through the primate amygdala. J Neurosci. Aug 28 2013;33(35):14017-14030.

15. Hubel DH, Wiesel TN. The period of susceptibility to the physiological effects of unilateral eye closure in kittens. Journal of Physiology. 1970;206(2):419-436.

16. Fagiolini M, Hensch TK. Inhibitory threshold for critical-period activation in primary visual cortex. Nature. 2000;404:183-186.

17. Hensch TK, Fagiolini M, Mataga N, Stryker MP, Baekkeskov S, Kash SF, Wiesel TN, Hubel DH, Olson CR, Freeman RD, Shatz CJ, Stryker MP, Antonini A, Stryker MP, Shatz CJ, Kratz KE, Spear PD, Smith DC, Duffy FH, Snodgrass SR, Burchfiel JR, Conway JL, Blakemore C, Hawken MJ, Sillito AM, Kemp JA, Blakemore C, Mower GD, Christen WG, Reiter HO, Stryker MP, Hata Y, Stryker MP, Ramoa AS, Paradiso MA, Freeman RD, Videen TO, Daw NW, Collins RC, Erlander MG, Tillakaratne NJK, Feldblum S, Patel N, Tobin AJ, Kaufman DL, Houser CR, Tobin AJ, Asada H, Guo Y, Kaplan IV, Cooper NGF, Mower GD, Fukuda T, Aika Y, Heizmann CW, Kosaka T, Reetz A, Martin DL, Martin SB, Wu SJ, Espina N, Asada H, Kash SF, Mataga N, Imamura K, Watanabe Y, Greif KF, Erlander MG, Tillakaratne NJK, Tobin AJ, Greif KF, Kaczmarek L, Chaudhuri A, Worley PF, Mataga N, Gordon JA, Stryker MP, Rauschecker JP, Tsumoto T, Kirkwood A, Rioult MG, Bear MF, Singer W, Katz LC, Shatz CJ, Study RE, Barker JL, Tallman JF, Gallagher DW, MacDonald RL, Olsen RW, Rogers CJ, Twyman RE, MacDonald RL, Eghbali M, Curmi JP, Birnir B, Gage PW, Shaw C, Aoki C, Wilkinson M, Prusky G, Cynader M, Xiang Z, Huguenard JR, Prince DA, Sigel E, Buhr A, Costa E, Berrueta LA, Gallo B, Vicente F, Zafra F, Castren E, Thoenen H, Lindholm D, Thoenen H, McAllister AK, Lo DC, Katz LC, Bonhoeffer T,

Sala R, Rutherford LC, Nelson SB, Turrigiano GG, Artola A, Singer W, Bear MF, Kirkwood A, Dudek SM, Friedlander MJ, Miller KD, Hensch TK, Stryker MP, Hensch TK, DeFelipe J, Gonchar Y, Burkhalter A, Kang Y, Kaneko T, Ohishi H, Endo K, Araki T, Brederode JFMV, Spain WJ, Komatsu Y, Carder RK, Leclerc SS, Hendry SHC, Kawaguchi Y, Parra D, Gulyas AI, Miles R, Kawaguchi Y, Shindou T, Xiang Z, Huguenard JR, Prince DA, Tamas G, Somogyi P, Buhl E, Gulyas A, Miles R, Hajos N, Freund TF, Buhl E, Halasy K, Somogyi P, Soltesz I, Smetters DK, Mody I, Miles R, Tamas G, Buhl EH, Somogyi P, Kawaguchi Y, Kubota Y, Benardo LS, Kawaguchi Y, Sugita S, Johnson SW, North RA, Segal M. Local GABA circuit control of experience-dependent plasticity in developing visual cortex. Science (New York, N.Y.). 1998;282:1504-1508.

18. Kuhl PK, Andruski JE, Chistovich IA, Chistovich LA, Kozhevnikova EV, Ryskina VL, Stolyarova EI, Sundberg U, Lacerda F. Cross-language analysis of phonetic units in language addressed to infants. Science. Aug 1 1997;277(5326):684-686.

19. Werker JF, Gilbert JH, Humphrey K, Tees RC. Developmental aspects of cross-language speech perception. Child Dev. Mar 1981;52(1):349-355.

20. Pascalis O, Scott LS, Kelly DJ, Shannon RW, Nicholson E, Coleman M, Nelson CA. Plasticity of face processing in infancy. Proc Natl Acad Sci U S A. Apr 5 2005;102(14):5297-5300.

21. Casey BJ, Trainor RJ, Orendi JL, Schubert AB, Nystrom LE, Giedd JN, Castellanos FX, Haxby JV, Noll DC, Cohen JD, Forman SD, Dahl RE, Rapoport JL. A Developmental functional MRI study of prefrontal activation during performance of a go-no-go task. Journal of Cognitive Neuroscience. 1997;9(6):835-847.

22. Gogtay N, Giedd JN, Lusk L, Hayashi KM, Greenstein D, Vaituzis AC, Nugent TF, 3rd, Herman DH, Clasen LS, Toga AW, Rapoport JL, Thompson PM. Dynamic mapping of human cortical development during childhood through early adulthood. Proc Natl Acad Sci U S A. May 25 2004;101(21):8174-8179.

23. Sowell ER, Thompson PM, Holmes CJ, Jernigan TL, Toga AW. In vivo evidence for post-adolescent brain maturation in frontal and striatal regions. Nature Neuroscience. 1999;2(10):859-861.

24. Durston S, Davidson MC, Tottenham N, Galvan A, Spicer J, Fossella JA, Casey BJ. A shift from diffuse to focal cortical activity with development. Dev Sci. Jan 2006;9(1):1-8.

25. Huttenlocher P. Synaptogenesis, synapse elimination, and neural plasticity in human cerbral cortex. In: Nelson C, ed. Threats to Optimal Development: Integrating Biological, Psychological, ad Social Risk Factors. Vol 27. New Jersy: Erlbaum; 1994:35-54.

26. Bourgeois JP. Synaptogenesis, heterochrony and epigenesis in the mammalian neocortex. Acta Paediatr Suppl. Jul 1997;422:27-33.

27. Gee DG, Humphreys KL, Flannery J, Goff B, Telzer EH, Shapiro M, Hare TA, Bookheimer SY, Tottenham N. A developmental shift from positive to negative connectivity in human amygdala-prefrontal circuitry. J Neurosci. Mar 6 2013;33(10):4584-4593.

28. Casey BJ. Beyond simple models of self-control to circuit-based accounts of adolescent behavior. Annu Rev Psychol. Jan 3 2015;66:295-319.

29. Silvers JA, Insel C, Powers A, Franz P, Helion C, Martin R, Weber J, Mischel W, Casey BJ, Ochsner KN. The transition from childhood to adolescence is marked

by a general decrease in amygdala reactivity and an affect-specific ventral-to-dorsal shift in medial prefrontal recruitment. Dev Cogn Neurosci. Jul 2 2016.

30. Galván A, Hare T, Parra C, Penn J, Voss H, Glover G, Casey B. Earlier development of the accumbens relative to orbitofrontal cortex might underlie risk-taking behavior in adolescents. Journal of Neuroscience. 2006;26(25):6885-6892.

31. Vink M, Derks JM, Hoogendam JM, Hillegers M, Kahn RS. Functional differences in emotion processing during adolescence and early adulthood. Neuroimage. May 2014;91:70-76.

32. Decety J, Michalska KJ, Kinzler KD. The contribution of emotion and cognition to moral sensitivity: a neurodevelopmental study. Cereb Cortex. Jan 2012;22(1):209-220.

33. Swartz JR, Carrasco M, Wiggins JL, Thomason ME, Monk CS. Age-related changes in the structure and function of prefrontal cortex-amygdala circuitry in children and adolescents: a multi-modal imaging approach. Neuroimage. Feb 2014;86:212-220.

34. Gabard-Durnam LJ, Flannery J, Goff B, Gee DG, Humphreys KL, Telzer E, Hare T, Tottenham N. The development of human amygdala functional connectivity at rest from 4 to 23 years: A cross-sectional study. Neuroimage. Mar 2014;95C:193-207.

35. Gabard-Durnam LJ, Gee DG, Goff B, Flannery J, Telzer E, Humphreys KL, Lumian DS, Fareri DS, Caldera C, Tottenham N. Stimulus-Elicited Connectivity Influences Resting-State Connectivity Years Later in Human Development: A Prospective Study. J Neurosci. Apr 27 2016;36(17):4771-4784.

36. Perlman SB, Pelphrey KA. Developing connections for affective regulation: age-related changes in emotional brain connectivity. J Exp Child Psychol. Mar 2011;108(3):607-620.

37. Silvers JA, Insel C, Powers A, Franz P, Helion C, Martin RE, Weber J, Mischel W, Casey BJ, Ochsner KN. vlPFC-vmPFC-Amygdala Interactions Underlie Age-Related Differences in Cognitive Regulation of Emotion. Cereb Cortex. Jun 23 2016.

38. Wu M, Kujawa A, Lu LH, Fitzgerald DA, Klumpp H, Fitzgerald KD, Monk CS, Phan KL. Age-related changes in amygdala-frontal connectivity during emotional face processing from childhood into young adulthood. Hum Brain Mapp. May 2016;37(5):1684-1695.

39. Motzkin JC, Philippi CL, Wolf RC, Baskaya MK, Koenigs M. Ventromedial prefrontal cortex is critical for the regulation of amygdala activity in humans. Biol Psychiatry. Feb 1 2015;77(3):276-284.

40. Ducharme S, Albaugh MD, Hudziak JJ, Botteron KN, Nguyen TV, Truong C, Evans AC, Karama S, Brain Development Cooperative G. Anxious/depressed symptoms are linked to right ventromedial prefrontal cortical thickness maturation in healthy children and young adults. Cereb Cortex. Nov 2014;24(11):2941-2950.

41. Dougherty LR, Blankenship SL, Spechler PA, Padmala S, Pessoa L. An fMRI Pilot Study of Cognitive Reappraisal in Children: Divergent Effects on Brain and Behavior. J Psychopathol Behav Assess. Dec 01 2015;37(4):634-644.

42. Yang EJ, Lin EW, Hensch TK. Critical period for acoustic preference in mice. Proc Natl Acad Sci U S A. Oct 2012;109 Suppl 2:17213-17220.

43. Arruda-Carvalho M, Wu W, Cummings KA, Clem R. Optogenetic examination of prefrontal-amygdala synaptic development. Journal of Neuroscience. in press.

44. Cressman VL, Balaban J, Steinfeld S, Shemyakin A, Graham P, Parisot N, Moore H. Prefrontal cortical inputs to the basal amygdala undergo pruning during late adolescence in the rat. J Comp Neurol. Jul 15 2010;518(14):2693-2709.

45. Bouwmeester H, Smits K, Van Ree JM. Neonatal development of projections to the basolateral amygdala from prefrontal and thalamic structures in rat. J Comp Neurol. Aug 26 2002;450(3):241-255.

46. Bouwmeester H, Wolterink G, van Ree JM. Neonatal development of projections from the basolateral amygdala to prefrontal, striatal, and thalamic structures in the rat. J Comp Neurol. Jan 14 2002;442(3):239-249.

47. Kim JH, Hamlin AS, Richardson R. Fear extinction across development: the involvement of the medial prefrontal cortex as assessed by temporary inactivation and immunohistochemistry. J Neurosci. Sep 2 2009;29(35):10802-10808.

48. Silvers JA, Lumian DS, Gabard-Durnam L, Gee DG, Goff B, Fareri DS, Caldera C, Flannery J, Telzer EH, Humphreys KL, Tottenham N. Previous Institutionalization Is Followed by Broader Amygdala-Hippocampal-PFC Network Connectivity during Aversive Learning in Human Development. J Neurosci. Jun 15 2016;36(24):6420-6430.

49. Campos JJ. Human emotions: Their new importance and their role in social referencing. Research & Clinical Center for Child Development. 1981;Annual Rpt.:1-7.

50. Walden TA, Ogan TA. The development of social referencing. Child Dev. Oct 1988;59(5):1230-1240.

51. Zarbatany L, Lamb ME. Social referencing as a function of information source: Mothers versus strangers. Infant Behavior and Development. 1985;8(1):25-33.

52. Aktar E, Majdandzic M, de Vente W, Bogels SM. Parental social anxiety disorder prospectively predicts toddlers' fear/avoidance in a social referencing paradigm. J Child Psychol Psychiatry. Jan 2014;55(1):77-87.

53. de Rosnay M, Cooper PJ, Tsigaras N, Murray L. Transmission of social anxiety from mother to infant: an experimental study using a social referencing paradigm. Behav Res Ther. Aug 2006;44(8):1165-1175.

54. Hostinar CE, Sullivan RM, Gunnar MR. Psychobiological mechanisms underlying the social buffering of the hypothalamic-pituitary-adrenocortical axis: a review of animal models and human studies across development. Psychol Bull. Jan 2014;140(1):256-282.

55. Gee DG, Gabard-Durnam L, Telzer EH, Humphreys KL, Goff B, Shapiro M, Flannery J, Lumian DS, Fareri DS, Caldera C, Tottenham N. Maternal buffering of human amygdala-prefrontal circuitry during childhood but not during adolescence. Psychol Sci. Nov 2014;25(11):2067-2078.

56. Howell BR, McMurray MS, Guzman DB, Nair G, Shi Y, McCormack KM, Hu X, Styner MA, Sanchez MM. Maternal buffering beyond glucocorticoids: impact of early life stress on corticolimbic circuits that control infant responses to novelty. Soc Neurosci. Jun 27 2016:1-15.

57. Kikusui T, Winslow JT, Mori Y. Social buffering: relief from stress and anxiety. Philos Trans R Soc Lond B Biol Sci. Dec 29 2006;361(1476):2215-2228.

58. Hennessy MB, Kaiser S, Sachser N. Social buffering of the stress response: diversity, mechanisms, and functions. Front Neuroendocrinol. Oct 2009;30(4):470-482.

59. Hostinar CE, Johnson AE, Gunnar MR. Parent support is less effective in buffer-

ing cortisol stress reactivity for adolescents compared to children. Dev Sci. Mar 2015;18(2):281-297.

60. Moriceau S, Sullivan RM. Maternal presence serves as a switch between learning fear and attraction in infancy. Nat Neurosci. Aug 2006;9(8):1004-1006.

61. Shionoya K, Moriceau S, Bradstock P, Sullivan RM. Maternal attenuation of hypothalamic paraventricular nucleus norepinephrine switches avoidance learning to preference learning in preweanling rat pups. Horm Behav. Sep 2007;52(3):391-400.

62. Debiec J, Sullivan RM. Intergenerational transmission of emotional trauma through amygdala-dependent mother-to-infant transfer of specific fear. Proc Natl Acad Sci U S A. Aug 19 2014;111(33):12222-12227.

63. Callaghan BL, Tottenham N. The Neuro-Environmental Loop of Plasticity: A Cross-Species Analysis of Parental Effects on Emotion Circuitry Development Following Typical and Adverse Caregiving. Neuropsychopharmacology. Jan 2016;41(1):163-176.

64. Callaghan BL, Tottenham N. The Stress Acceleration Hypothesis: effects of early-life adversity on emotion circuits and behavior. Current Opinion in Behavioral Sciences. 2016.

8
Olfaction: Smell of Change in the Air

1. Buck L and Axel R (1991) A novel multigene family may encode odorant receptors: a molecular basis for odor recognition. Cell 65 (1):175-187

2. Greer PL, Bear DM, Lassance JM, Bloom ML, Tsukahara T, Pashkovski SL, Masuda FK, Nowlan AC, Kirchner R, Hoekstra HE, and Datta SR (2016) A Family of non-GPCR chemosensors defines an alternative logic for mammalian olfaction. Cell 165 (7):1734-1748

3. Busse D, Kudella P, Gruning NM, Gisselmann G, Stander S, Luger T, Jacobsen F, Steinstrasser L, Paus R, Gkogkolou P, Bohm M, Hatt H, and Benecke H (2014) A synthetic sandalwood odorant induces wound-healing processes in human keratinocytes via the olfactory receptor OR2AT4. J Invest Dermatol 134 (11):2823-2832

4. Pavlath GK (2010) A new function for odorant receptors: MOR23 is necessary for normal tissue repair in skeletal muscle. Cell Adh Migr 4 (4):502-506

5. Kang N, Kim H, Jae Y, Lee N, Ku CR, Margolis F, Lee EJ, Bahk YY, Kim MS, and Koo J (2015) Olfactory marker protein expression is an indicator of olfactory receptor-associated events in non-olfactory tissues. PLoS One 10 (1):e0116097

6. Foster SR, Roura E, Molenaar P, and Thomas WG (2015) G protein-coupled receptors in cardiac biology: old and new receptors. Biophys Rev 7 (1):77-89

7. Trotier D, Bensimon JL, Herman P, Tran Ba HP, Doving KB, and Eloit C (2007) Inflammatory obstruction of the olfactory clefts and olfactory loss in humans: a new syndrome? Chem Senses 32 (3):285-292

8. Kang N, Bahk YY, Lee N, Jae Y, Cho YH, Ku CR, Byun Y, Lee EJ, Kim MS, and Koo J (2015) Olfactory receptor Olfr544 responding to azelaic acid regulates glucagon secretion in alpha-cells of mouse pancreatic islets. Biochem Biophys Res Commun 460 (3):616-621

9. Kudyakova TI, Sarycheva NY, and Kamenskii AA (2007) Orientation and ex-

ploratory behavior and anxiety of CBA mice with anosmia induced by N-trimethylindole (skatole). Bull Exp Biol Med 143 (1):1-4

10. Massberg D, Jovancevic N, Offermann A, Simon A, Baniahmad A, Perner S, Pungsrinont T, Luko K, Philippou S, Ubrig B, Heiland M, Weber L, Altmuller J, Becker C, Gisselmann G, Gelis L, and Hatt H (2016) The activation of OR51E1 causes growth suppression of human prostate cancer cells. Oncotarget 7 (30):48231-48249

11. Shepard BD, Cheval L, Peterlin Z, Firestein S, Koepsell H, Doucet A, and Pluznick JL (2016) A Renal Olfactory Receptor Aids in Kidney Glucose Handling. Sci Rep 6:35215

12. Braun T, Voland P, Kunz L, Prinz C, and Gratzl M (2007) Enterochromaffin cells of the human gut: sensors for spices and odorants. Gastroenterology 132 (5):1890-1901

− 13. Weber M, Pehl U, Breer H, and Strotmann J (2002) Olfactory receptor expressed in ganglia of the autonomic nervous system. J Neurosci Res 68 (2):176-184

14. Chen Z, Zhao H, Fu N, and Chen L (2017) The diversified function and potential therapy of ectopic olfactory receptors in non-olfactory tissues. J Cell Physiol

15. Air Force Instructions 48-123. Accession, Retention and Administration. Medical Examination and Standards, Volume 2, 5 June 2006.

16. Doty, R. L. Handbook of Olfaction and Gustation. 3rd, 1-1217. 2015. Hoboken, N.J., John Wiley & Sons, Inc.

17. Doty RL, Shaman P, Applebaum SL, Giberson R, Siksorski L, and Rosenberg L (1984) Smell identification ability: changes with age. Science 226:1441-1443

18. Doty RL, Shaman P, and Dann M (1984) Development of the University of Pennsylvania Smell Identification Test: a standardized microencapsulated test of olfactory function. Physiology & Behavior 32:489-502

19. Doty RL (1995) The Smell Identification TestTM Administration Manual -- 3rd Edition. Sensonics, Inc., Haddon Hts., NJ

20. Stevens JC and Cain WS (1986) Aging Impairs the Ability to Perceive Gas Odor. Chemical Senses 11 (4):668

21. Pence TS, Reiter ER, DiNardo LJ, and Costanzo RM (2014) Risk factors for hazardous events in olfactory-impaired patients. JAMA Otolaryngol Head Neck Surg 140 (10):951-955

22. Devanand DP, Lee S, Manly J, Andrews H, Schupf N, Masurkar A, Stern Y, Mayeux R, and Doty RL (2015) Olfactory identification deficits and increased mortality in the community. Ann Neurol 78 (3):401-411

23. Pinto JM, Wroblewski KE, Kern DW, Schumm LP, and McClintock MK (2014) Olfactory dysfunction predicts 5-year mortality in older adults. PLoS One 9 (10):e107541

24. Doty RL, Hawkes CH, Good KP, and Duda JE (2015) Odor perception and neuropathology in neurodegenerative diseases and schizophrenia. In: Doty RL (ed) Handbook of Olfaction and Gustation. John Wiley & Sons, Hoboken, pp 403-452

25. Braak H, Braak E, Yilmazer D, de Vos RAI, Jansen ENH, and Bohl J (1996) Pattern of brain destruction in Parkinson's and Alzheimer's diseases. J Neural Trans 103:455-490

26. Doty RL (2017) Olfactory dysfunction in neurodegenerative diseases: is there a common pathological substrate? Lancet neurol 16 (6):478-488

27. Sohrabi HR, Bates KA, Rodrigues M, Taddei K, Laws SM, Lautenschlager NT, Dhaliwal SS, Johnston AN, Mackay-Sim A, Gandy S, Foster JK, and Martins RN (2009) Olfactory dysfunction is associated with subjective memory complaints in community-dwelling elderly individuals. J Alzheimers Dis 17 (1):135-142

28. Devanand DP, Tabert MH, Cuasay K, Manly JJ, Schupf N, Brickman AM, Andrews H, Brown TR, DeCarli C, and Mayeux R (2010) Olfactory identification deficits and MCI in a multi-ethnic elderly community sample. Neurobiol Aging 31 (9):1593-1600

29. Devanand DP, Lee S, Manly J, Andrews H, Schupf N, Doty RL, Stern Y, Zahodne LB, Louis ED, and Mayeux R (2015) Olfactory deficits predict cognitive decline and Alzheimer dementia in an urban community. Neurology 84 (2):182-189

30. Graves AB, Bowen JD, Rajaram L, McCormick WC, McCurry SM, Schellenberg GD, and Larson EB (1999) Impaired olfaction as a marker for cognitive decline: interaction with apolipoprotein E epsilon4 status. Neurology 53 (7):1480-1487

31. Serby M, Mohan C, Aryan M, Williams L, Mohs RC, and Davis KL (1996) Olfactory identification deficits in relatives of Alzheimer's disease patients. Biol Psychiat 39 (5):375-377

32. Calderon-Garciduenas L, Franco-Lira M, Henriquez-Roldan C, Osnaya N, Gonzalez-Maciel A, Reynoso-Robles R, Villarreal-Calderon R, Herritt L, Brooks D, Keefe S, Palacios-Moreno J, Villarreal-Calderon R, Torres-Jardon R, Medina-Cortina H, Delgado-Chavez R, Aiello-Mora M, Maronpot RR, and Doty RL (2010) Urban air pollution: Influences on olfactory function and pathology in exposed children and young adults. Experimental and Toxicological Pathology 62:91-102

33. Tabert MH, Liu X, Doty RL, Serby M, Zamora D, Pelton GH, Marder K, Albers MW, Stern Y, and Devanand DP (2005) A 10-item smell identification scale related to risk for Alzheimer's disease. Ann Neurol 58 (1):155-160

34. Ross GW, Petrovitch H, Abbott RD, Tanner CM, Popper J, Masaki K, Launer L, and White LR (2008) Association of olfactory dysfunction with risk for future Parkinson's disease. Ann Neurol 63 (2):167-173

35. Ponsen MM, Stoffers D, Wolters EC, Booij J, and Berendse HW (2010) Olfactory testing combined with dopamine transporter imaging as a method to detect prodromal Parkinson's disease. J Neurol Neurosurg Psychiatry 81 (4):396-399

36. Deeb J, Shah M, Muhammed N, Gunasekera R, Gannon K, Findley LJ, and Hawkes CH (2010) A basic smell test is as sensitive as a dopamine transporter scan: comparison of olfaction, taste and DaTSCAN in the diagnosis of Parkinson's disease. Quarterly Journal of Medicine 103 (12):941-952

37. Schofield PW, Ebrahimi H, Jones AL, Bateman GA, and Murray SR (2012) An olfactory 'stress test' may detect preclinical Alzheimer's disease. BMC Neurol 12:24

38. Doty RL, Golbe LI, McKeown DA, Stern MB, Lehrach CM, and Crawford D (1993) Olfactory testing differentiates between progressive supranuclear palsy and idiopathic Parkinson's disease. Neurology 43 (5):962-965

39. Solomon GS, Petrie WM, Hart JR, and Brackin HB, Jr. (1998) Olfactory dysfunction discriminates Alzheimer's dementia from major depression. Journal of Neuropsychiatry & Clinical Neurosciences 1998 Winter;10 (1):64-67

40. Suchowersky O, Reich S, Perlmutter J, Zesiewicz T, Gronseth G, and Weiner WJ

(2006) Practice Parameter: Diagnosis and prognosis of new onset Parkinson disease (an evidence-based review) Report of the Quality Standards Subcommittee of the American Academy of Neurology. Neurology 66 (7):968-975

41. Salihoglu M, Kendirli MT, Altundag A, Tekeli H, Saglam M, Cayonu M, Senol MG, and Ozdag F (2014) The effect of obstructive sleep apnea on olfactory functions. Laryngoscope 124 (9):2190-2194

42. Walliczek-Dworschak U, Cassel W, Mittendorf L, Pellegrino R, Koehler U, Guldner C, Dworschak POG, Hildebrandt O, Daniel H, Gunzel T, Teymoortash A, and Hummel T (2017) Continuous positive air pressure improves orthonasal olfactory function of patients with obstructive sleep apnea. Sleep Med 34:24-29

43. Schubert CR, Fischer ME, Pinto AA, Klein BEK, Klein R, and Cruickshanks KJ (2017) Odor detection thresholds in a population of older adults. Laryngoscope 127 (6):1257-1262

44. Rosenfeldt AB, Dey T, and Alberts JL (2016) Aerobic Exercise Preserves Olfaction Function in Individuals with Parkinson's Disease. Parkinsons Dis 2016:9725089

45. Gopinath B, Sue CM, Flood VM, Burlutsky G, and Mitchell P (2015) Dietary intakes of fats, fish and nuts and olfactory impairment in older adults. Br J Nutr 114 (2):240-247

46. Stevenson RJ, Boakes RA, Oaten MJ, Yeomans MR, Mahmut M, and Francis HM (2016) Chemosensory Abilities in Consumers of a Western-Style Diet. Chem Senses 41 (6):505-513

47. Duffy VB, Backstrand JR, and Ferris AM (1995) Olfactory dysfunction and related nutritional risk in free-living, elderly women. J Am Diet Assoc 95 (8):879-884

48. Siderowf A, Jennings D, Connolly J, Doty RL, Marek K, and Stern MB (2007) Risk factors for Parkinson's disease and impaired olfaction in relatives of patients with Parkinson's disease. Mov Disord 22 (15):2249-2255

49. Menco BPM and Morrison EE (2003) Morphology of the mammalian olfactory epithelium: form, fine structure, function, and pathology. In: Doty RL (ed) Handbook of olfaction and gustation. Marcel Dekker, New York, pp 17-49

50. Fusar-Poli P, Rubia K, Rossi G, Sartori G, and Balottin U (2012) Striatal dopamine transporter alterations in ADHD: pathophysiology or adaptation to psychostimulants? A meta-analysis. Am J Psychiatry 169 (3):264-272

9
The Illusion of the Perfect Brain Enhancer

1 Antal A, Alekseichuk I, Bikson M, Brockmöller J, Brunoni AR, Chen R et al. Low intensity transcranial electric stimulation: Safety, ethical, legal regulatory and application guidelines. Clin Neurophysiol 2017. doi:10.1016/j.clinph.2017.06.001.

2 Philip NS, Nelson BG, Frohlich F, Lim KO, Widge AS, Carpenter LL. Low-Intensity Transcranial Current Stimulation in Psychiatry. Am J Psychiatry 2017; 174: 628–639.

3 Adventures in Transcranial Direct-Current Stimulation. New Yorker. http://www.newyorker.com/magazine/2015/04/06/electrified (accessed 17 Jul2017).

4 Stagg CJ, Nitsche MA. Physiological basis of transcranial direct current stimulation. Neuroscientist 2011; 17: 37–53.

5 Pascual-Leone A, Pridmore H. Transcranial magnetic stimulation (TMS). Aust-NZJ Psychiatry 1995; 29: 698.

6   Olympians zap their brains for Rio 2016. MIMS News. https://today.mims.com/topic/olympians-zap-their-brains-with-transcranial-direct-current-stimulation-for-rio-2016 (accessed 17 Jul2017).

7   For the Golden State Warriors, Brain Zapping Could Provide an Edge. New Yorker. http://www.newyorker.com/tech/elements/for-the-golden-state-warriors-brain-zapping-could-provide-an-edge (accessed 17 Jul2017).

8   Katwala A. How the brain could be the next doping battleground. The Telegraph. http://www.telegraph.co.uk/news/2016/08/09/neural-doping---how-the-brain-could-be-the-next-doping-battlegro/ (accessed 17 Jul2017).

9   Leinenga G, Langton C, Nisbet R, Gotz J. Ultrasound treatment of neurological diseases - current and emerging applications. NatRevNeurol 2016; 12: 161–174.

10 Pascual-Leone A, Walsh V. Transcranial magnetic stimulation. MIT Press 2005; 29: 698.

11 Antal A, Boros K, Poreisz C, Chaieb L, Terney D, Paulus W. Comparatively weak after-effects of transcranial alternating current stimulation (tACS) on cortical excitability in humans. Brain Stimul 2008; 1: 97–105.

12 Nitsche MA, Paulus W. Excitability changes induced in the human motor cortex by weak transcranial direct current stimulation. J Physiol 2000; 527 Pt 3: 633–639.

13 Kim H, Chiu A, Lee SD, Fischer K, Yoo SS. Focused ultrasound-mediated non-invasive brain stimulation: examination of sonication parameters. Brain Stimul 2014; 7: 748–756.

14 Yoo SS, Bystritsky A, Lee JH, Zhang Y, Fischer K, Min BK et al. Focused ultrasound modulates region-specific brain activity. Neuroimage 2011; 56: 1267–1275.

15 Santarnecchi E, Brem AK, Levenbaum E, Thompson T, Kadosh RC, Pascual-Leone A. Enhancing cognition using transcranial electrical stimulation. Curr Opin Behav Sci 2015; 171–178.

16 Rotenberg A, Horvath JC, Pascual-Leone A. Transcranial Magnetic Stimulation (Neuromethods). Springer Sci Bus Media NY 2014; 22: 257–266.

17 Walsh V, Pascual-Leone A. Neurochronometrics of Mind: Transcranial magnetic stimulation in Cognitive Science. MIT Press 2003; 29: 698.

18 Pascual-Leone A, Gates JR, Dhuna A. Induction of speech arrest and counting errors with rapid-rate transcranial magnetic stimulation. Neurology 1991; 41: 697–702.

19 Pascual-Leone A, Catala MD, Pascual-Leone PA. Lateralized effect of rapid-rate transcranial magnetic stimulation of the prefrontal cortex on mood. Neurology 1996; 46: 499–502.

20 Fecteau S, Knoch D, Fregni F, Sultani N, Boggio P, Pascual-Leone A. Diminishing risk-taking behavior by modulating activity in the prefrontal cortex: a direct current stimulation study. J Neurosci 2007; 27: 12500–12505.

21 Knoch D, Pascual-Leone A, Meyer K, Treyer V, Fehr E. Diminishing Reciprocal Fairness by Disrupting the Right Prefrontal Cortex. Science 2006; 314: 829–832.

22 Young L, Camprodon JA, Hauser M, Pascual-Leone A, Saxe R. Disruption of the right temporoparietal junction with transcranial magnetic stimulation reduces the role of beliefs in moral judgments. ProcNatlAcadSciUSA 2010; 107: 6753–6758.

23 Fox MD, Halko MA, Eldaief MC, Pascual-Leone A. Measuring and manipulating

brain connectivity with resting state functional connectivity magnetic resonance imaging (fcMRI) and transcranial magnetic stimulation (TMS). Neuroimage 2012; 62: 2232–2243.

24 Halko MA, Farzan F, Eldaief MC, Schmahmann JD, Pascual-Leone A. Intermittent theta-burst stimulation of the lateral cerebellum increases functional connectivity of the default network. J Neurosci 2014; 34: 12049–12056.

25 Wang JX, Rogers LM, Gross EZ, Ryals AJ, Dokucu ME, Brandstatt KL et al. Targeted enhancement of cortical-hippocampal brain networks and associative memory. Science 2014; 345: 1054–1057.

26 Massimini M, Boly M, Casali A, Rosanova M, Tononi G. A perturbational approach for evaluating the brain's capacity for consciousness. ProgBrain Res 2009; 177: 201–214.

27 Shafi MM, Brandon WM, Oberman L, Cash SS, Pascual-Leone A. Modulation of EEG functional connectivity networks in subjects undergoing repetitive transcranial magnetic stimulation. Brain Topogr 2014; 27: 172–191.

28 Rosanova M, Casali A, Bellina V, Resta F, Mariotti M, Massimini M. Natural frequencies of human corticothalamic circuits. JNeurosci 2009; 29: 7679–7685.

29 Vernet M, Bashir S, Yoo WK, Perez JM, Najib U, Pascual-Leone A. Insights on the neural basis of motor plasticity induced by theta burst stimulation from TMS-EEG. EurJ Neurosci 2013; 37: 598–606.

30 Santarnecchi, Polizzotto NR, Godone M, Giovannelli F, Feurra M, Matzen L et al. Frequency-Dependent Enhancement of Fluid Intelligence Induced by Transcranial Oscillatory Potentials. Curr Biol 2013; 23: 1449–1453.

31 Lustenberger C, Boyle MR, Foulser AA, Mellin JM, Frohlich F. Functional role of frontal alpha oscillations in creativity. Cortex 2015; 67: 74–82.

32 Priori A, Berardelli A, Rona S, Accornero N, Manfredi M. Polarization of the human motor cortex through the scalp. Neuroreport 1998; 9: 2257–2260.

33 Buch ER, Santarnecchi E, Antal A, Born J, Celnik PA, Classen J et al. Effects of tDCS on motor learning and memory formation: A consensus and critical position paper. Clin Neurophysiol 2017; 128: 589–603.

34 Landhuis E, Landhuis E. Do DIY Brain-Booster Devices Work? Sci. Am. 2017. https://www.scientificamerican.com/article/do-diy-brain-booster-devices-work/ (accessed 11 Aug2017).

35 Transcranial Direct Current Stimulation • r/tDCS. reddit. https://www.reddit.com/r/tDCS/ (accessed 11 Aug2017).

36 Santarnecchi E, Feurra M, Galli G, Rossi A, Rossi S. Overclock your brain for gaming? Ethical, social and health care risks. Brain Stimul 2013; 6: 713–714.

37 Wurzman R, Hamilton RH, Pascual-Leone A, Fox MD. An open letter concerning do-it-yourself users of transcranial direct current stimulation: DIY-tDCS. Ann Neurol 2016; 80: 1–4.

38 Wexler A. The practices of do-it-yourself brain stimulation: implications for ethical considerations and regulatory proposals. J Med Ethics 2016; 42: 211–215.

39 Fitz NS, Reiner PB. The challenge of crafting policy for do-it-yourself brain stimulation. J Med Ethics 2015; 41: 410–412.

40 Maslen H, Douglas T, Cohen KR, Levy N, Savulescu J. The regulation of cognitive enhancement devices: extending the medical model. JLaw Biosci 2014; 1: 68–93.

41 Maslen H, Douglas T, Kadosh RC, Levy N, Savulescu J. Do-it-yourself brain

stimulation: a regulatory model. J Med Ethics 2013. doi:10.1136/medeth-ics-2013-101692.

42   Edelman G. Gerald Edelman: From Brain Dynamics to Consciousness: A Prelude to the Future of Brain-Based Devices. Brain Dyn. Conscious. Prelude Future Brain-Based Devices. 2007.https://www.youtube.com/watch?v=8mvH-Q6hLTLs (accessed 11 Aug2017).

43   Iuculano T, Cohen KR. The mental cost of cognitive enhancement. JNeurosci 2013; 33: 4482–4486.

44   Snowball A, Tachtsidis I, Popescu T, Thompson J, Delazer M, Zamarian L et al. Long-Term Enhancement of Brain Function and Cognition Using Cognitive Training and Brain Stimulation. CurrBiol 2013. doi:10.1016/j.cub.2013.04.045.

45   López-Alonso V, Cheeran B, Río-Rodríguez D, Fernández-del-Olmo M. Inter-individual Variability in Response to Non-invasive Brain Stimulation Paradigms. Brain Stimulat 2014; 7: 372–380.

46   Krause B, Cohen KR. Not all brains are created equal: the relevance of individual differences in responsiveness to transcranial electrical stimulation. Front SystNeurosci 2014; 8: 25.

47   Steenbergen L, Sellaro R, Hommel B, Lindenberger U, Kühn S, Colzato LS. 'Unfocus' on foc.us: commercial tDCS headset impairs working memory. Exp Brain Res 2016; 234: 637–643.

48   Maslen H, Savulescu J, Douglas T, Levy N, Cohen KR. Regulation of devices for cognitive enhancement. Lancet 2013; 382: 938–939.

49   McLellan TM, Caldwell JA, Lieberman HR. A review of caffeine's effects on cognitive, physical and occupational performance. Neurosci Biobehav Rev 2016; 71: 294–312.

50   LaCroix AZ, Mead LA, Liang K-Y, Thomas CB, Pearson TA. Coffee Consumption and the Incidence of Coronary Heart Disease. N Engl J Med 1986; 315: 977–982.

11
Microglia: The Brain's First Responders

1.   Bilbo, S.D. & Schwarz, J.M. (2012). The immune system and developmental programming of brain and behavior. Front Neuroendocrinol, 33(3), 267-286.

2.   Deverman, B.E. & Patterson, P.H. (2009). Cytokines and CNS development. Neuron, 64(1), 61-78.

3.   Lacagnina, M.J., Rivera, P.D., & Bilbo, S.D. (2017). Glial and Neuroimmune Mechanisms as Critical Modulators of Drug Use and Abuse. Neuropsychopharmacology, 42(1), 156-177.

4.   Sierra, A., de Castro, F., Del Rio-Hortega, J., Rafael Iglesias-Rozas, J., Garrosa, M., & Kettenmann, H. (2016). The "Big-Bang" for modern glial biology: Translation and comments on Pio del Rio-Hortega 1919 series of papers on microglia. Glia, 64(11), 1801-1840.

5.   Tremblay, M.E., Lecours, C., Samson, L., Sanchez-Zafra, V., & Sierra, A. (2015). From the Cajal alumni Achucarro and Rio-Hortega to the rediscovery of never-resting microglia. Front Neuroanat, 9, 45.

6.   Ginhoux, F., Greter, M., Leboeuf, M., Nandi, S., See, P., Gokhan, S., ... Merad, M. (2010). Fate mapping analysis reveals that adult microglia derive from primitive macrophages. Science, 330(6005), 841-845.

7. Lawson, L.J., Perry, V.H., Dri, P., & Gordon, S. (1990). Heterogeneity in the distribution and morphology of microglia in the normal adult mouse brain. Neuroscience, 39(1), 151-170.

8. Kierdorf, K. & Prinz, M. (2013). Factors regulating microglia activation. Front Cell Neurosci, 7, 44.

9. Hanisch, U.K. (2002). Microglia as a source and target of cytokines. Glia, 40(2), 140-155.

10. Lucin, K.M. & Wyss-Coray, T. (2009). Immune activation in brain aging and neurodegeneration: too much or too little? Neuron, 64(1), 110-122.

11. Ransohoff, R.M. & Perry, V.H. (2009). Microglial physiology: unique stimuli, specialized responses. Annual Review of Immunology, 27(Journal Article), 119-145.

12. Saijo, K. & Glass, C.K. (2011). Microglial cell origin and phenotypes in health and disease. Nat Rev Immunol, 11(11), 775-787.

13. D'Agostino, P.M., Gottfried-Blackmore, A., Anandasabapathy, N., & Bulloch, K. (2012). Brain dendritic cells: biology and pathology. Acta Neuropathol, 124(5), 599-614.

14. Kettenmann, H., Hanisch, U.K., Noda, M., & Verkhratsky, A. (2011). Physiology of microglia. Physiological Reviews, 91(2), 461-553.

15. Kreutzberg, G.W. (1996). Microglia: a sensor for pathological events in the CNS. Trends in neurosciences, 19(8), 312-318.

16. Stence, N., Waite, M., & Dailey, M.E. (2001). Dynamics of microglial activation: a confocal time-lapse analysis in hippocampal slices. Glia, 33(3), 256-266.

17. Davalos, D., Grutzendler, J., Yang, G., Kim, J.V., Zuo, Y., Jung, S., . . . Gan, W.B. (2005). ATP mediates rapid microglial response to local brain injury in vivo. Nat Neurosci, 8(6), 752-758.

18. Jung, S., Aliberti, J., Graemmel, P., Sunshine, M.J., Kreutzberg, G.W., Sher, A., & Littman, D.R. (2000). Analysis of fractalkine receptor CX(3)CR1 function by targeted deletion and green fluorescent protein reporter gene insertion. Mol Cell Biol, 20(11), 4106-4114.

19. Nimmerjahn, A., Kirchhoff, F., & Helmchen, F. (2005). Resting microglial cells are highly dynamic surveillants of brain parenchyma in vivo. Science, 308(5726), 1314-1318.

20. Hong, S., Beja-Glasser, V.F., Nfonoyim, B.M., Frouin, A., Li, S., Ramakrishnan, S., . . . Stevens, B. (2016). Complement and microglia mediate early synapse loss in Alzheimer mouse models. Science, 352(6286), 712-716.

21. Graeber, M.B. (2010). Changing face of microglia. Science (New York, N.Y.), 330(6005), 783-788.

22. Schafer, D.P., Lehrman, E.K., Kautzman, A.G., Koyama, R., Mardinly, A.R., Yamasaki, R., . . . Stevens, B. (2012). Microglia sculpt postnatal neural circuits in an activity and complement-dependent manner. Neuron, 74(4), 691-705.

23. Schafer, D.P., Lehrman, E.K., & Stevens, B. (2013). The "quad-partite" synapse: microglia-synapse interactions in the developing and mature CNS. Glia, 61(1), 24-36.

24. Tremblay, M.E. & Majewska, A.K. (2011). A role for microglia in synaptic plasticity? Communicative & integrative biology, 4(2), 220-222.

25. Tremblay, M.E., Lowery, R.L., & Majewska, A.K. (2010). Microglial interactions with synapses are modulated by visual experience. PLoS biology, 8(11), e1000527.

26. Wake, H., Moorhouse, A.J., Jinno, S., Kohsaka, S., & Nabekura, J. (2009). Resting microglia directly monitor the functional state of synapses in vivo and determine the fate of ischemic terminals. J Neurosci, 29(13), 3974-3980.

27. Schafer, D.P. & Stevens, B. (2013). Phagocytic glial cells: sculpting synaptic circuits in the developing nervous system. Curr Opin Neurobiol, 23(6), 1034-1040.

28. Boulanger, L.M. (2009). Immune Proteins in Brain Development and Synaptic Plasticity. Neuron, 64(1), 93-109.

29. Shatz, C.J. (2009). MHC class I: an unexpected role in neuronal plasticity. Neuron, 64(1), 40-45.

30. Stephan, A.H., Barres, B.A., & Stevens, B. (2012). The complement system: an unexpected role in synaptic pruning during development and disease. Annu Rev Neurosci, 35, 369-389.

31. Stevens, B., Allen, N.J., Vazquez, L.E., Howell, G.R., Christopherson, K.S., Nouri, N., . . . Barres, B.A. (2007). The classical complement cascade mediates CNS synapse elimination. Cell, 131(6), 1164-1178.

32. Chu, Y., Jin, X., Parada, I., Pesic, A., Stevens, B., Barres, B., & Prince, D.A. (2010). Enhanced synaptic connectivity and epilepsy in C1q knockout mice. Proc Natl Acad Sci U S A, 107(17), 7975-7980.

33. Paolicelli, R.C., Bolasco, G., Pagani, F., Maggi, L., Scianni, M., Panzanelli, P., . . . Gross, C.T. (2011). Synaptic pruning by microglia is necessary for normal brain development. Science, 333(6048), 1456-1458.

34. Sekar, A., Bialas, A.R., de Rivera, H., Davis, A., Hammond, T.R., Kamitaki, N., . . . McCarroll, S.A. (2016). Schizophrenia risk from complex variation of complement component 4. Nature, 530(7589), 177-183.

35. Morgan, J.T., Chana, G., Pardo, C.A., Achim, C., Semendeferi, K., Buckwalter, J., . . . Everall, I.P. (2010). Microglial activation and increased microglial density observed in the dorsolateral prefrontal cortex in autism. Biol Psychiatry, 68(4), 368-376.

36. Pardo, C.A., Vargas, D.L., & Zimmerman, A.W. (2005). Immunity, neuroglia and neuroinflammation in autism. Int Rev Psychiatry, 17(6), 485-495.

37. Vargas, D.L., Nascimbene, C., Krishnan, C., Zimmerman, A.W., & Pardo, C.A. (2005). Neuroglial activation and neuroinflammation in the brain of patients with autism. Ann Neurol, 57(1), 67-81.

38. Bland, S.T., Beckley, J.T., Young, S., Tsang, V., Watkins, L.R., Maier, S.F., & Bilbo, S.D. (2010). Enduring consequences of early-life infection on glial and neural cell genesis within cognitive regions of the brain. Brain Behav Immun, 24(3), 329-338.

39. Williamson, L.L., Sholar, P.W., Mistry, R.S., Smith, S.H., & Bilbo, S.D. (2011). Microglia and memory: modulation by early-life infection. J Neurosci, 31(43), 15511-15521.

40. Bilbo, S.D., Yirmiya, R., Amat, J., Paul, E.D., Watkins, L.R., & Maier, S.F. (2008). Bacterial infection early in life protects against stressor-induced depressive-like symptoms in adult rats. Psychoneuroendocrinology, 33(3), 261-269.

41. Bilbo, S.D., Biedenkapp, J.C., Der-Avakian, A., Watkins, L.R., Rudy, J.W., & Maier, S.F. (2005). Neonatal infection-induced memory impairment after lipopolysaccharide in adulthood is prevented via caspase-1 inhibition. J Neurosci, 25(35), 8000-8009.

42. Reu, P., Khosravi, A., Bernard, S., Mold, J.E., Salehpour, M., Alkass, K., . . . Frisen,

J. (2017). The Lifespan and Turnover of Microglia in the Human Brain. Cell Rep, 20(4), 779-784.

43. Bilbo, S.D., Block, C.L., Bolton, J.L., Hanamsagar, R., & Tran, P.K. (2017). Beyond infection - Maternal immune activation by environmental factors, microglial development, and relevance for autism spectrum disorders. Exp Neurol.

44. Ransohoff, R.M. (2016). How neuroinflammation contributes to neurodegeneration. Science, 353(6301), 777-783.

45. Efthymiou, A.G. & Goate, A.M. (2017). Late onset Alzheimer's disease genetics implicates microglial pathways in disease risk. Mol Neurodegener, 12(1), 43.

46. Salter, M.W. & Stevens, B. (2017). Microglia emerge as central players in brain disease. Nat Med, 23(9), 1018-1027.

47. Hong, S. & Stevens, B. (2016). Microglia: Phagocytosing to Clear, Sculpt, and Eliminate. Dev Cell, 38(2), 126-128.

48. Shi, Q., Chowdhury, S., Ma, R., Le, K.X., Hong, S., Caldarone, B.J., . . . Lemere, C.A. (2017). Complement C3 deficiency protects against neurodegeneration in aged plaque-rich APP/PS1 mice. Sci Transl Med, 9(392).

49. DeKosky, S.T., Scheff, S.W., & Styren, S.D. (1996). Structural correlates of cognition in dementia: quantification and assessment of synapse change. Neurodegeneration, 5(4), 417-421.

50. Terry, R.D., Masliah, E., Salmon, D.P., Butters, N., DeTeresa, R., Hill, R., . . . Katzman, R. (1991). Physical basis of cognitive alterations in Alzheimer's disease: synapse loss is the major correlate of cognitive impairment. Ann Neurol, 30(4), 572-580.

51. Lui, H., Zhang, J., Makinson, S.R., Cahill, M.K., Kelley, K.W., Huang, H.Y., . . . Huang, E.J. (2016). Progranulin Deficiency Promotes Circuit-Specific Synaptic Pruning by Microglia via Complement Activation. Cell, 165(4), 921-935.

52. Howell, G.R., Macalinao, D.G., Sousa, G.L., Walden, M., Soto, I., Kneeland, S.C., . . . John, S.W. (2011). Molecular clustering identifies complement and endothelin induction as early events in a mouse model of glaucoma. J Clin Invest, 121(4), 1429-1444.

53. Hanamsagar, R. & Bilbo, S.D. (2017). Environment matters: microglia function and dysfunction in a changing world. Curr Opin Neurobiol, 47, 146-155.

12

Neuroimaging Advances for Depression

1. Ostergaard SD, Jensen SO, Bech P. The heterogeneity of the depressive syndrome: when numbers get serious. Acta Psychiatr Scand 2011; 124: 495-496.

2. American Psychiatric Association: Practice Guideline for the Treatment of Patients with Major Depressive Disorder. 3rd Ed. American Psychiatric Association. Washington DC, 2010.

3. Lewis AJ. Melancholia: a clinical survey of depressive states. J Mental Sci 1934; 80: 277-378.

4. Kiloh LG, Andrews G, Neilson M, Bianchi GN (1972). The relationship of the syndromes called endogenous and neurotic depression. British Journal of Psychiatry 121: 183.

5. Horwitz AV, Wakefield JC. The Loss of Sadness: How Psychiatry Transformed Normal Sorrow into Depressive Disorder. Oxford University Press. New York, 2007.

6. American Psychiatric Association. Diagnostic and statistical manual of mental disorders: DSM-5 (5th ed.). Arlington, VA: American Psychiatric Association. Arlington, VA, 2013. .

7. Pae CU, Tharwani H, Marks DM, Masand PS, Patkar AA. Atypical depression: a comprehensive review. CNS Drugs 2009; 23: 1023-1037.

8. Perlis RH. A clinical risk stratification tool for predicting treatment resistance in major depressive disorder. Biol Psychiatry 2013; 74: 7 14.

9. Wallace ML, Frank E, Kraemer HC. A novel approach for developing and interpreting treatment moderator profiles in randomized clinical trials. JAMA Psychiatry 2013; 70: 1241-1247.

10. DeRubeis RJ, Cohen ZD, Forand NR, Fournier JC, Gelfand LA, Lorenzo-Luaces L. The Personalized Advantage Index: translating research on prediction into individualized treatment recommendations. A demonstration. PLoS One 2014; 9: e83875.

11. Huibers MJ, Cohen ZD, Lemmens LH, Arntz A, Peeters FP, Cuijpers P, et al. Predicting optimal outcomes in cognitive therapy or interpersonal psychotherapy for depressed individuals using the Personalized Advantage Index approach. PLoS One 2015; 10: e0140771.

12. Jha MK, Minhajuddin A, Gadad BS, Greer T, Grannemann B, Soyombo A, et al. Can C-reactive protein inform antidepressant medication selection in depressed outpatients? Findings from the CO-MED trial. Psychoneuroendocrinology 2017; 78: 105-113.

13. Uher R, Tansey KE, Dew T, Maier W, Mors O, Hauser J, et al. An inflammatory biomarker as a differential predictor of outcome of depression treatment with escitalopram and nortriptyline. Am J Psychiatry 2014; 171: 1278-1286.

14. Raison CL, Rutherford RE, Woolwine BJ, Shuo C, Schettler P, Drake DF, et al. A randomized controlled trial of the tumor necrosis factor antagonist infliximab for treatment-resistant depression: the role of baseline inflammatory biomarkers. JAMA Psychiatry 2013; 70: 31-41.

15. Savitz JB, Rauch SL, Drevets WC. Clinical application of brain imaging for the diagnosis of mood disorders: the current state of play. Mol Psychiatry 2013; 18: 528-539.

16. Savitz JB, Drevets WC. Neuroreceptor imaging in depression. Neurobiol Dis 2013; 52: 49-65.

17. Strawbridge R, Young AH, Cleare AJ. Biomarkers for depression: recent insights, current challenges and future prospects. Neuropsychiatric Dis Treat 2017; 13: 1245-1262.

18. Kambeitz J, Cabral C, Sacchet MD, Gotlib IH, Zahn R, Serpa MH, et al. Detecting neuroimaging biomarkers for depression: A meta-analysis of multivariate pattern recognition studies. Biol Psychiatry 2017; 82: 330-338.

19. Ressler KJ, Mayberg HS. Targeting abnormal neural circuits in mood and anxiety disorders: from the laboratory to the clinic. Nat Neurosci 2007; 10: 1116-1124.

20. Kaiser RH, Andrews-Hanna JR, Wager TD, Pizzagalli DA. Large-scale network dysfunction in major depressive disorder: A meta-analysis of resting-state functional connectivity. JAMA Psychiatry 2015; 72: 603-611.

21. Schmaal L, Veltman DJ, van Erp TG, Sämann PG, Frodl T, Jahanshad N, et al. Subcortical brain alterations in major depressive disorder: findings from the ENIGMA Major Depressive Disorder working group. Mol Psychiatry 2016; 21: 806-812.

22. Lorenzetti V, Allen NB, Fornito A, Yücel M. Structural brain abnormalities in major depressive disorder: a selective review of recent MRI studies. J Affect Disord 2009; 117: 1-17.

23. Schmaal L, Hibar DP, Sämann PG, Hall GB, Baune BT, Jahanshad N, et al. Cortical abnormalities in adults and adolescents with major depression based on brain scans from 20 cohorts worldwide in the ENIGMA Major Depressive Disorder Working Group. Mol Psychiatry 2017; 22: 900-909.

24. Colle R, Dupong I, Colliot O, Deflesselle E, Hardy P, Falissard B, et al. Smaller hippocampal volumes predict lower antidepressant response/remission rates in depressed patients: A meta-analysis. World J Biol Psychiatry 2016; 15: 1-8.

25. Fu CH, Steiner H, Costafreda SG. Predictive neural biomarkers of clinical response in depression: a meta-analysis of functional and structural neuroimaging studies of pharmacological and psychological therapies. Neurobiol Dis 2013; 52: 75-83.

26. Gillihan SJ, Parens E. Should we expect "neural signatures" for DSM diagnoses? J Clin Psychiatry 2011; 72: 1383-1389.

27. Dunlop BW. How shall I diagnose thee? Let me count the ways. Biol Psychiatry 2017; 82: 306-308.

28. Kessler RC, Chiu WT, Demler O, Merikangas KR, Walters EE. Prevalence, severity, and comorbidity of 12-month DSM-IV disorders in the National Comorbidity Survey Replication. Arch Gen Psychiatry 2005; 62: 617-627.

29. Rakofsky JJ, Dunlop BW. The over-under on misdiagnosis of bipolar disorder: A systematic review. Curr Psychiatry Rev 2015; 11: 222-234.

30. Rive MM, Redlich R, Schmaal L, Marquand AF, Dannlowski U, Grotegerd D, et al. Distinguishing medication-free subjects with unipolar disorder from subjects with bipolar disorder: state matters. Bipolar Disord 2016; 18: 612-623.

31. Redlich R, Almeida JJ, Grotegerd D, Opel N, Kugel H, Heindel W, et al. Brain morphometric biomarkers distinguishing unipolar and bipolar depression. A voxel-based morphometry-pattern classification approach. JAMA Psychiatry 2014; 71: 1222-1230.

32. Ambrosi E, Arciniegas DB, Madan A, Curtis KN, Patriquin MA, Jorge RE, et al. Insula and amygdala resting-state functional connectivity differentiate bipolar from unipolar depression. Acta Psychiatr Scand 2017; 136: 129-139.

33. Bürger C, Redlich R, Grotegerd D, Meinert S, Dohm K, Schneider I, et al. Differential abnormal pattern of anterior cingulate gyrus activation in unipolar and bipolar depression: an fMRI and pattern classification approach. Neuropsychopharmacol 2017; 42: 1399-1408.

34. De Almeida JRC, Phillips ML. Distinguishing between unipolar depression and bipolar depression: Current and future clinical and neuroimaging perspectives. Biol Psychiatry 2013; 73: 111-118.

35. Phillips ML, Kupfer DJ. Bipolar disorder diagnosis: challenges and future directions. Lancet 2013; 381(9878): 1663-1671.

36. Williams LM. Precision psychiatry: a neural circuit taxonomy for depression and anxiety. Lancet Psychiatry 2016; 3: 472-480.

37. Vassilopoulou K, Papathanasiou M, Michopoulos I, Boufidou F, Oulis P, Kelekis N, et al. A magnetic resonance imaging study of hippocampal, amygdala and subgenual prefrontal cortex volumes in major depression subtypes: melancholic versus psychotic depression. J Affect Disord 2013; 146: 197-204.

38. Bracht T, Horn H, Strik W, Federspiel A, Schnell S, Höfle O, Stegmayer K, Wiest R, Dierks T, Müller TJ, Walther S. White matter microstructure alterations of the medial forebrain bundle in melancholic depression. J Affect Disord 2014; 155: 186-193.

39. Ota M, Noda T, Sato N, Hattori K, Hori H, Sasayama D, et al. White matter abnormalities in major depressive disorder with melancholic and atypical features: A diffusion tensor imaging study. Psychiatry Clin Neurosci 2015; 69: 360-368.

40. Greenberg DL, Payne ME, MacFall JR, Steffens DC, Krishnan RR. Hippocampal volumes and depression subtypes. Psychiatry Res 2008; 163: 126-132.

41. Feder S, Sundermann B, Wersching H, Teuber A, Kugel H, Teismann H, et al. Sample heterogeneity in unipolar depression as assessed by functional connectivity analyses is dominated by general disease effects. J Affect Disord 2017; 222: 79-87.

42. Guo CC, Hyett MP, Nguyen VT, Parker GB, Breakspear MJ. Distinct neurobiological signatures of brain connectivity in depression subtypes during natural viewing of emotionally salient films. Psychol Med 2016; 46: 1535-1545.

43. Hyett MP, Breakspear MJ, Friston KJ, Guo CC, Parker GB. Disrupted effective connectivity of cortical systems supporting attention and interoception in melancholia. JAMA Psychiatry 2015; 72: 350-358.

44. Stringaris A, Vidal-Ribas Belil P, Artiges E, Lemaitre H, Gollier-Briant F, Wolke S, et al. The brain's response to reward anticipation and depression in adolescence: dimensionality, specificity, and longitudinal predictions in a community-based sample. Am J Psychiatry 2015; 172: 1215-1223.

45. Nelson BD, Perlman G, Klein DN, Kotov R, Hajcak G. Blunted neural response to rewards as a prospective predictor of the development of depression in adolescent girls. Am J Psychiatry 2016; 173: 1223-1230.

46. Müller VI, Cieslik EC, Serbanescu I, Laird AR, Fox PT, Eickhoff SB. Altered brain activity in unipolar depression revisited: Meta-analyses of neuroimaging studies. JAMA Psychiatry 2017; 74: 47-55.

47. Siegle GJ, Thompson W, Carter CS, Steinhauer SR, Thase ME. Increased amygdala and decreased dorsolateral prefrontal BOLD responses in unipolar depression: related and independent features. Biol Psychiatry 2007; 61: 198-209.

48. Palmer SM, Crewther SG, Carey LM; START Project Team. A meta-analysis of changes in brain activity in clinical depression. Front Hum Neurosci 2015; 8: 1045.

49. Zeng LL, Shen H, Liu L, Hu D. Unsupervised classification of major depression using functional connectivity MRI. Hum Brain Mapp 2014; 35: 1630-1641.

50. Drysdale AT, Grosenick L, Downar J, Dunlop K, Mansouri F, Meng Y, et al. Resting-state connectivity biomarkers define neurophysiological subtypes of depression. Nat Med 2017; 23: 28-38.

51. Williams LM, Korgaonkar MS, Song YC, Paton R, Eagles S, Goldstein-Piekarski A, et al. Amygdala reactivity to emotional faces in the prediction of general and medication-specific responses to antidepressant treatment in the randomized iSPOT-D trial. Neuropsychopharmacol 2015; 40: 2398-2408.

52. Goldstein-Piekarski AN, Korgaonkar MS, Green E, Suppes T, Schatzberg AF, Hastie T, et al. Human amygdala engagement moderated by early life stress exposure is a biobehavioral target for predicting recovery on antidepressants. Proc Natl Acad Sci USA 2016; 113: 11955-11960.

53. Crane NA, Jenkins LM, Bhaumik R, Dion C, Gowins JR, Mickey BJ, et al. Multidimensional prediction of treatment response to antidepressants with cognitive control and functional MRI. Brain 2017; 140: 472-486.

54. Siegle GJ, Thompson WK, Collier A, Berman SR, Feldmiller J, Thase ME, et al. Toward clinically useful neuroimaging in depression treatment: prognostic utility of subgenual cingulate activity for determining depression outcome in cognitive therapy across studies, scanners, and patient characteristics. Arch Gen Psychiatry 2012; 69: 913-924.

55. Crowther A, Smoski MJ, Minkel J, Moore T, Gibbs D, Petty C, et al. Resting-state connectivity predictors of response to psychotherapy in major depressive disorder. Neuropsychopharmacol 2015; 40: 1659-1673.

56. Carl H, Walsh E, Eisenlohr-Moul T, Minkel J, Crowther A, Moore T, et al. Sustained anterior cingulate cortex activation during reward processing predicts response to psychotherapy in major depressive disorder. J Affect Disord 2016; 203: 204-212.

57. Dichter GS, Gibbs D, Smoski MJ. A systematic review of relations between resting-state functional-MRI and treatment response in major depressive disorder. J Affect Disord 2015; 172: 8-17.

58. McGrath CL, Kelley ME, Holtzheimer PE, Dunlop BW, Craighead WE, Franco AR, et al. Toward a neuroimaging treatment selection biomarker for major depressive disorder. JAMA Psychiatry 2013; 70: 821-829.

59. Dunlop BW, Kelley ME, McGrath CL, Craighead WE, Mayberg HS. Preliminary findings supporting insula metabolic activity as a predictor of outcome to psychotherapy and medication treatments for depression. J Neuropsychiatry Clin Neurosci 2015; 27: 237-239.

60. Dunlop BW, Rajendra JK, Craighead WE, Kelley, ME, McGrath CL, Choi KS, et al. Functional connectivity of the subcallosal cingulate cortex identifies differential outcomes to treatment with cognitive behavior therapy or antidepressant medication for major depressive disorder. Am J Psychiatry 2017; 174: 533-545.

61. McGrath CL, Kelley ME, Dunlop BW, Holtzheimer PE, Craighead WE, Mayberg HS. Pretreatment brain states identify likely nonresponse to standard treatments for depression. Biol Psychiatry 2014; 76: 527-535

Index